Practice Papers for SQA Exams

Standard Grade | Credit

Biology

ISBN 978-1-84372-765-1

Published by
Leckie & Leckie Ltd, 4 Queen Street, Edinburgh, EH2 1JE
Tel: 0131 220 6831 Fax: 0131 225 9987
enquiries@leckieandleckie.co.uk   www.leckieandleckie.co.uk

For Emilio

A CIP Catalogue record for this book is available from the British Library.

Leckie & Leckie Ltd is a division of Huveaux plc.

Questions and answers in this book do not emanate from SQA. All of our entirely new and original Practice Papers have been written by experienced authors working directly for the publisher.

Introduction

Layout of the Book

The papers which follow have been produced to give you practice in two of the three elements needed for Standard Grade Biology. The first element is "Knowledge and Understanding" [KU] which depends on you being very familiar with all the theory, especially important in Biology which is very much a knowledge-based subject. You need to know the facts, concepts and common techniques. The second element is "Problem Solving" [PS] which tests your ability to handle data, do calculations, draw and/or interpret graphs, extract information from diagrams, read passages, draw conclusions, predict what will happen. These two elements, which are assessed by the formal National Examination, are combined with your internal "Practical Assessment" done in school to produce your overall grade.

All four papers have been modelled on the actual National Examination with a balance of KU and PS, each element worth 40 marks making 80 in total. All parts of the syllabus at Credit Level have been represented in these papers. The written examination will be worth 80% of your final grade award with the remainder coming from the internal assessment.

The layout, paper colour and question level of all four papers are all very similar to the actual exam that you will sit, so that you are familiar with what the exam paper will look like. In addition, the format of the questions reflects those you are likely to encounter in the National Examination, giving you practice at handling these.

The answer section is at the back of the book. Each answer contains, where appropriate, some guidance as to how this was obtained, practical tips on how to tackle certain types of questions, details of how marks are awarded and advice on just what the examiners will be looking for.

How to use This Book

Interacting with learning material is a powerful means of obtaining feedback on your strengths and weaknesses. The material in these papers will give you an excellent way of working on different strategies needed to handle the actual National Examination. Seeking help, where needed, from your teacher is vital to improvement and building up your confidence and expertise.

The Practice Papers can be used in two main ways:

1. You can complete an entire practice paper as preparation for the final exam. If you would like to use the book in this way, you can either complete the practice paper under exam style conditions by setting yourself a time for each paper and answering it as well as possible without using any references or notes. Alternatively, you can answer the practice paper questions as a revision exercise, using your notes to produce a model answer. Your teacher may mark these for you.

2. You can use the Topic Index at the front of this book to find all the questions within the book that deal with a specific topic. This allows you to focus specifically on areas that you particularly want to revise or, if you are mid-way through your course, it lets you practise answering exam-style questions for just those topics that you have studied. You will find this useful if you want practice on handling particular types of problem-solving questions.

Revision Advice

The need to work out a plan for regular and methodical revision is obvious. If you leave things to the last minute, it may result in panic and stress which will inhibit you from performing to your maximum. If you need help, it is best to find this out when there is time to put it right. Revision planners are highly individual and you need to produce one which suits you. Use an area of your home which is set aside only for studying if possible so that you form a positive link and in this way be less liable to distractions. Not only do you need a plan for revising Biology, but also for all your subjects. Below is one version but you will have your own ideas here!

Work out a revision timetable for each week's work in advance – remember to cover all of your subjects and to leave time for homework and breaks. For example:

Day	6pm–6.45pm	7pm–8pm	8.15pm–9pm	9.15pm–10pm
Monday	Homework	Homework	English Revision	Biology Revision
Tuesday	Maths Revision	Physics Revision	Homework	Free
Wednesday	Geography Revision	Modern Studies Revision	English Revision	French Revision
Thursday	Homework	Maths Revision	Biology Revision	Free
Friday	Geography Revision	French Revision	Free	Free
Saturday	Free	Free	Free	Free
Sunday	Modern Studies Revision	Maths Revision	Modern Studies Revision	Homework

Make sure that you have at least one evening free a week to relax, socialise and re-charge your batteries. It also gives your brain a chance to process the information that you have been feeding it all week.

Arrange your study time into one hour or 30 minute sessions, with a break between sessions e.g. 6pm–6.45pm, 7pm–8pm, 8.15pm–9pm. Try to start studying as early as possible in the evening when your brain is still alert and be aware that the longer you put off starting, the harder it will be to start!

If you miss a session, don't panic. Log this and make it up as soon as possible. Don't get behind in your schedule. Discipline is everything in being a successful student.

Study a different subject in each session, except for the day before an exam.

Do something different during your breaks between study sessions – have a cup of tea, or listen to some music. Don't let your 15 minutes expand into 20 or 25 minutes though!

Have your class notes and any textbooks available for your revision to hand as well as plenty of blank paper, a pen, etc. You may like to make keyword sheets like the example below:

Keyword	Meaning
Cell	Basic unit of living things
ADH	Hormone which regulates water balance
Epidermis	Layer of cells on upper or lower surface of leaves

"Flash cards" are an excellent way of practising terms and definitions. You can make these easily or buy them very cheaply. Use the the flash cards either to recall the keyword when you see the meaning or to give the meaning when you see the keyword.

Finally forget or ignore all or some of the advice in this section if you are happy with your present way of studying. Everyone revises differently, so find a way that works for you!

Command Words

In the practice papers and in the National Examination itself, a number of "command words" will be used in the questions. These command words are used to show how you should answer a question – some words indicate that you should write more than others. If you familiarise yourself with these command words, it will help you to structure your answers more effectively.

Command Word	Meaning/Explanation
Name, state, identify, list	Giving a list is acceptable here – as a general rule you will get one mark for each point you give
Suggest	Give more than a list – perhaps a proposal or an idea
Outline	Give a brief description or overview of what you are talking about
Describe	Give more detail than you would in an outline, and use examples where you can
Explain	Discuss why an action has been taken or an outcome reached – what are the reasons and/or processes behind it

Justify	Give reasons for your answer, stating why you have taken an action or reached a particular conclusion
Define	Give the meaning of the term
Compare	Give the key features of 2 different items or ideas and discuss their similarities and/or their differences
Predict	Work out what will happen

In the Exam

Watch your time and pace yourself carefully. Work out roughly how much time you can spend on each answer and try to stick to this.

Be clear before the examination what the instructions are likely to be, e.g. how many questions you should answer in each section. The practice papers will help you to become familiar with the examination's instructions.

Read the question thoroughly before you begin to answer it – make sure you know exactly what the question is asking you to do. If the question is in sections, e.g. 15a, 15b, 15c, etc, make sure that you can answer each section before you start writing.

Don't repeat yourself as you will not get any more marks for saying the same thing twice. This also applies to annotated diagrams which will not get you any extra marks if the information is repeated in the written part of your answer.

Give proper explanations. A common error is to give descriptions rather than explanations. If you are asked to explain something, you should be giving reasons. Check your answer to an 'explain' question and make sure that you have used linking words and phrases such as 'because', 'this means that', 'therefore', 'so', 'so that', 'due to', 'since' and 'the reason is'.

Good luck!

Topic Index

Topic	Sub-Topic	Exam A	Exam B	Exam C	Exam D
1 THE BIOSPHERE	Investigating an ecosystem	1(a)	3		9(b)
	How it works	2		9	11(a), (b), (c)
	Control and management	1(b)			3
2 THE WORLD OF PLANTS	Introducing plants		1, 17		
	Growing plants	8(a), (b), (c)			7
	Making food	8(d), 15		15	
3 ANIMAL SURVIVAL	The need for food		5, 11(c)	11	
	Reproduction		7		16
	Water and waste	3		4	
	Responding to the environment			1	
4 INVESTIGATING CELLS	Investigating living cells		9(a)		
	Investigating diffusion	12(b)	9(b)		4
	Investigating cell division	10			
	Investigating enzymes				13
	Investigating aerobic respiration		11(a) (i)–(ii), (b), (d)		
5 THE BODY IN ACTION	Movement	11(c)	13		1
	The need for energy				
	Coordination	11(a), (b)			6
	Changing levels of performance			6	

6 INHERITANCE — Variation		15(a)		
What is inheritance?	6(a) (i)–(ii), (b)	15(d)		13(a), (c), (e)
Genetics and society		15(b)		
7 BIOTECHNOLOGY — Living factories				7(a), (b) (i), (v)
Problems and profit with waste	13(a), (b), (c)		9	9(a)
Reprogramming microbes	5(c)		14	
Selecting information	1(a), 4, 5(a), (b), 9(a) (ii), 16(a), (b), (c)	4, 8(a), (b), (c), 10(a), (c)	3, 6(c) (i), 10(b), (c), 12(a), (b), (c), (d)	2, 5(a), (c), (d), 12(a), 14(a), 15(b)
Presenting information	2(a) (i), 7(a), 9(a) (i), 13(d), 14(a)	2(a), 6(a), 12(a), (b), 16(b)	2(a), 5(b), 8(a), 10(a)	8(b), 10(a), 14(b), 15(a)
Calculations	6(a) (iii), 7(b), (c), 12(a), 14(b), (c), 16(d)	2(b), (c), 6(c), 11(a) (ii), 12(c), 14(a), (b), (d), 16(a), (c), (e)	2(b), (d), 5(a), 6(c) (ii), 8(b), 10(d), (e), 13(b), (d)	5(b), 8(a), 10(b) (i), 12(c), 15(d), (e)
Experimental procedures	5(d), 9(b)	10(d), 12(d)	2(c), 7(b) (iv), 10(f)	10(c), (d), 15(c)
Drawing conclusions/describing relationships	9(c)	6(b), 10(b), 12(c), 16(d)	6(c) (iii), 7(b) (ii), (iii)	5(e), 8(c), 10(b) (ii), 12(b), 14(c)
Making predictions		6(d)	8(c)	

Exam A – Credit Level

Biology

Standard Grade: Credit

Practice Papers
For SQA Exams

Exam A
Credit Level

Fill in these boxes:

Name of centre

Town

Forename(s)

Surname

Try to answer all of the questions in the time allowed.

You have 1 hour, 30 minutes to complete this paper.

Write your answers in the spaces provided, including all of your working.

1. (*a*) The following table lists identifiable features of some common invertebrate animals.

| Feature | Invertebrate animals | | | | |
	Scorpion	Earthworm	Ragworm	Crayfish	Ant
legs	eight present	absent	absent	ten present	six present
body	hard and segmented	soft and segmented	soft and unsegmented	hard and segmented	hard and segmented
habitat	lives on land	lives in soil	lives in sea	lives in sea	lives on land

(i) Using this information, complete the boxes in the paired statement key below.

1 Legs present .... go to 2

   ☐ .... go to 4

2 Lives in sea .... *crayfish*
   Does not live in sea .... go to 3

3 ☐ .... ☐

   More than six legs present .... ☐

4 Segments present .... *earthworm*
   Segments absent .... ☐

(ii) Give **one** similarity and **one** difference between a crayfish and a scorpion.

Similarity _____

_____

Difference _____

_____

(*b*) Crayfish are usually found in clean, unpolluted water. What are such animals, whose presence shows the state of the water, called?

_____

KU	PS
	3
	1
	1

	KU	PS

2. (*a*) (i) Use the information below to complete the food web shown.

Hognose snake eats toads

Hawks eat garter snakes, rabbits and hognose snakes

Grasshoppers are eaten by spiders, garter snakes and toads

Grass is eaten by grasshoppers and rabbits

Toads eat spiders

(ii) **Underline** and explain your prediction of the effect on the hognose snake population if the toads were removed from this web.

The population of hognose snakes would   increase

decrease

stay the same

Explanation _____

_____

_____

(*b*) As you move along a food chain, what usually is the relationship between the size of the organisms and the number of the organisms?

_____

_____

2 (PS column)

1 (KU column, near ii)

1 (KU column, near b)

	KU	PS

3. The diagram below shows one filter unit in the human kidney.

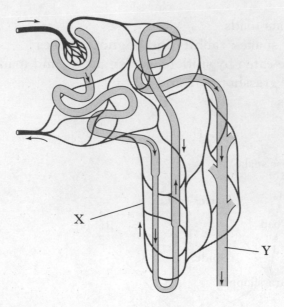

(a) (i) What is this unit called? _____   1

    (ii) Indicate, by means of a line and letter G, the position of the glomerulus.   1

    (iii) What is the function of each structure indicated by the letters X and Y?

        X _____

        _____

        Y _____

        _____   2

(b) How is the waste urea produced in the body?

_____   1

4. Read the following passage and then answer the questions that follow.

There is a common impression that no higher animal can live long without drinking water. Certainly this is true of humans and many other mammals; we need water at frequent intervals and in a very hot, dry desert a human without water cannot last more than a day or so. An animal such as the camel can survive somewhat longer but sooner or later it too must drink to refill its supply.

Yet we know that the waterless desert is not uninhabited. Even in desert areas with no visible drinking water within many miles, one will often find a fairly rich animal life. How do these animals get the water they must have to live? The body of a desert rat has about the same water content (65% of body weight) as that of a drinking animal and it generally has no more tolerance to drying out of its cells, sometimes less. For many desert animals the answer is simple; they get their water in their food. These animals live on juicy plants, one of the most important of which is cactus. The pack rat, for example, eats large quantities of cactus pulp, which is about 90% water. Thus it is easy to account for the survival of animals in areas where cacti and other water-storing plants are available.

There are, however, animals which can live in areas altogether barren of juicy vegetation. An outstanding example is a certain general type of desert rodent which is found in all the major desert areas of the world. The so-called kangaroo rat is not actually related to the kangaroo, though it looks a great deal like one. It hops along on long hind legs and it has a long, strong tail which it uses for support and steering. It lives in a burrow in the ground by day and comes out for food only at night. The animal does very well even in the driest areas. Water to drink, even dew, is rarely available in its natural habitat. The kangaroo rat has only a short range of movement – not more than a few hundred yards – and therefore does not leave its dry area to find juicy plants. Its food consists of seeds and other dry plant material.

Adapted from *The Desert Rat* by Knut and Bodil Schmidt-Nielsen in *Readings from Scientific American – From Cell to Organism.* published by W.H. Freeman and Company 1967. Library of Congress Catalogue Number 66–30156.

(*a*) How long would a human be able to survive in a very dry environment?

_____

(*b*) What is the water content of a desert rat?

_____

	KU	PS

(c) How do many desert animals obtain their water?

_____

_____ | | 1

(d) State **two** functions of the kangaroo rat's tail.

1 _____

2 _____ | | 1

(e) How does the behaviour of the kangaroo rat help conserve water?

_____

_____ | | 1

(f) Why does the kangaroo rat not move any distance from its normal habitat?

_____

_____ | | 1

(g) What does the kangaroo rat feed on?

_____

_____ | | 1

5. A person was suffering from a bad throat infection and a sample was taken and cultured on an agar plate which had a number of different antibiotics on small discs. The results are shown below.

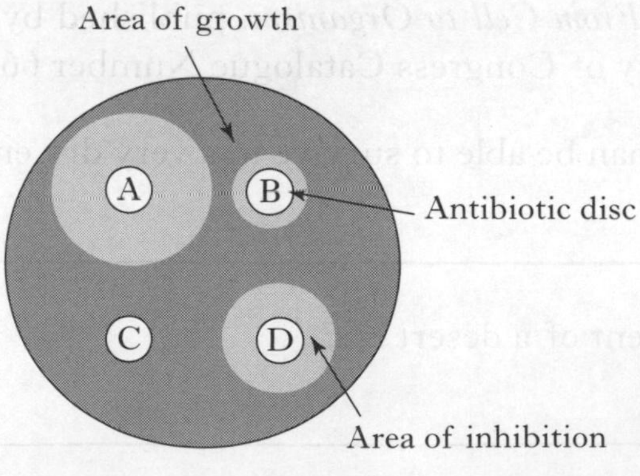

Area of growth

Antibiotic disc

Area of inhibition

(a) The person was known to be allergic to antibiotic A. Which antibiotic should be given? Explain your answer.

Antibiotic _____

Explanation _____

_____ | | 1

	KU	PS

(b) Which antibiotic is the least effective?

_____   1 (PS)

(c) Explain why a number of different antibiotics is needed to treat diseases caused by bacteria.

_____

_____   1 (KU)

(d) How could the reliability of this procedure be improved?

_____

_____   1 (PS)

6. The following cross was carried out using plants which have either orange or cream coloured fruits.

P   orange fruit   ×   cream fruit

F_1 all orange fruits

     F_1 × F_1

F_2 orange and cream fruits

(a) (i) Using the letter O for orange and o for cream coloured fruits, give the genotypes and phenotypes of the parents and the F_1 generation.

	Genotype	*Phenotype*
Parents		
F_1		

(ii) Predict the phenotypic and genotypic ratios for the F_2 generation.

Genotypic ratio   ____ : ____ : ____

Phenotypic ratio   _____ : _____   2

(iii) Predict the expected number of cream fruits if 500 orange fruits had been formed in the F_2 generation.

Space for calculation

_____ cream fruits   1 (PS)

(b) What are different forms of a gene called? _____   1 (KU)

7. The data below shows information about blood groups in a human population.

% Population with blood group			
A	B	AB	O
30	40	5	25

(a) Plot this data as a **bar chart** on the grid below adding appropriate labels.

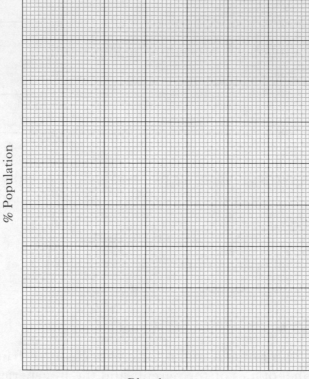

Blood groups

2

(b) In a town, 1200 people live. How many of them would you predict would be blood group AB?

Space for calculation

1

(c) Why do these results not allow an accurate description of the proportions of different blood groups in the human population as whole?

1

8. The diagram below represents an insect-pollinated flower.

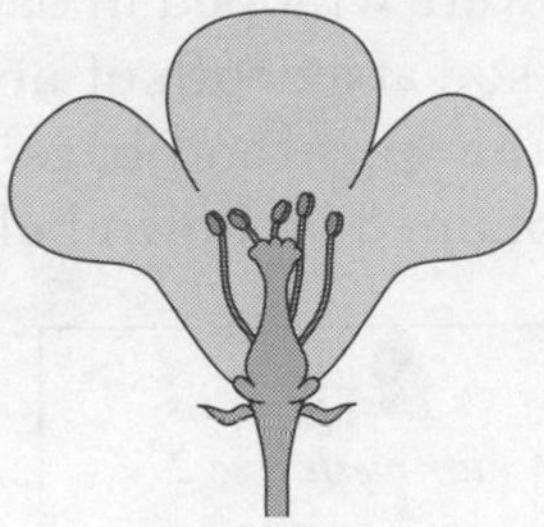

(a) Name **one** structure shown and explain how it is involved in insect pollination.

Structure _____

Explanation _____

_____

1

(b) State **two** differences between insect-pollinated and wind-pollinated flowers.

1 _____

_____

2 _____

_____

2

(c) What is the term used to describe plants which are genetically identical and produced asexually from a single parent?

_____

1

(d) Give **one** function of xylem in a plant.

_____

_____

1

KU	PS

KU	PS

9. An experiment was carried out to see how quickly some garden waste broke down in the soil. Identical masses of waste were wrapped in bags of two different mesh sizes, 1 and 2, and then buried in the soil at a depth of around 10cm. Mesh 1 was larger than mesh 2. Every 4 weeks, the waste and bags were removed, weighed and returned to the soil. The data obtained is shown below.

Time in soil (days)	% Loss of mass for mesh size 1	% Loss of mass for mesh size 2
0	0	0
28	16	10
56	60	12
84	72	28
112	80	36

(a) (i) Plot these data as two **line graphs** on the grid below.

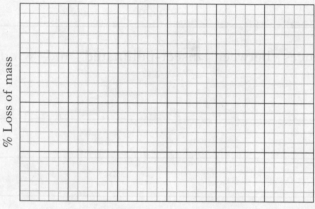

Time in soil (days)

2

(ii) Which mesh size, 1 or 2, produced the fastest **rate** of breakdown of the waste?

_____

1

(b) Suggest **one** source of error in the procedure adopted for this experiment.

_____

1

(c) State **one** conclusion you can make from the results.

1 _____

_____

1

10. The diagrams below show 4 stages in the process of mitosis.

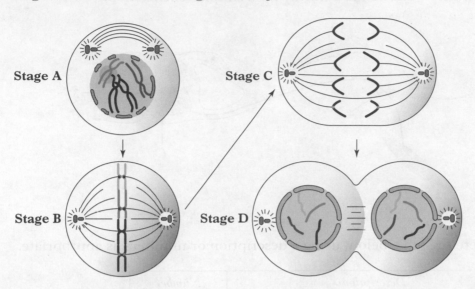

(*a*) Describe what is happening at stages B and C.

B _____

_____

C _____

_____

(*b*) The number of chromosomes is preserved in the new cells formed by mitosis. Explain the significance of this.

_____

_____

KU	PS
1	
1	
1	

11. The diagram below shows the pathway for a simple reflex action.

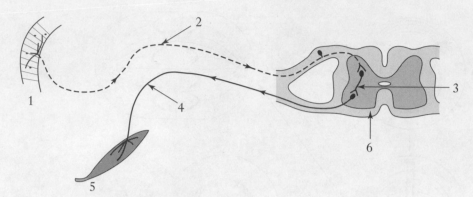

(a) Complete the table below, using a description or numbers as appropriate.

Description	Number
Organ which reacts to the stimulus	
	4
Sensory nerve cell	
Relay nerve cell	

(b) **Underline one** answer in each bracket to describe how the central nervous system deals with information.

Changes in the outside environment are picked up by the $\begin{pmatrix} \text{brain} \\ \text{sense organs} \\ \text{liver} \end{pmatrix}$

and transferred to the $\begin{pmatrix} \text{central nervous sytem} \\ \text{digestive system} \\ \text{circulatory system} \end{pmatrix}$ which may respond by

causing $\begin{pmatrix} \text{bones} \\ \text{muscles} \end{pmatrix}$ to react.

(c) Name the structures which attach bones to

muscles _____ bones _____

KU 2

KU 1

KU 2

12. The diagram below is a typical plant cell.

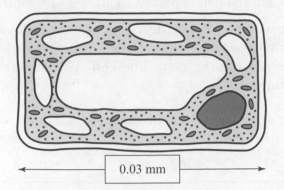

0.03 mm

The unit used to measure cell size is the micrometre (μm).

1000 micrometres = 1 mm

(a) (i) What is the length of this cell expressed in micrometres?

Space for calculation

_____ μm

1

(ii) If a human red blood cell's diameter is roughly one quarter of the length of this plant cell, calculate the diameter of the red blood cell in micrometres.

Space for calculation

_____ μm

1

(iii) A bacterial cell was found to be 5 micrometres in length. How many times longer is the plant cell?

Space for calculation

_____

1

(iv) Express the ratio of the plant, red blood and bacterial cell lengths.

____ : ____ : ____

plant cell   red cell   bacterial cell

1

KU	PS

(b) (i) **Underline one** word in each pair so that the description of what happens when an animal cell is placed in strong salt solution is correct.

Water moves [up/down] a concentration gradient as it enters a cell by osmosis. This is because the water is in [higher/lower] concentration outside relative to inside the cell. The cell will [shrink/burst] if this movement continues.

(ii) Other than water, state **one** other substance which must pass into animal cells for them to function.

13. (a) The treatment of sewage involves breakdown by micro-organisms such as bacteria. How do these organisms benefit from this breakdown?

(b) Why is the complete breakdown of sewage dependent on oxygen being present?

(c) Bacteria may be able to produce animal feed from industrial wastes.

State **one** way in which the food value of the animal feed has been increased.

	KU	PS
	1	
	1	
	1	
	1	
	1	

(d) The data below shows what was found in a typical "wheelie bin" from a home in Scotland.

Description	Percentage
Cans, aluminium foil	10
Plastic and glass	20
Garden waste	10
Paper and cardboard	20
Kitchen waste	40

Present this data as a pie-chart using the blank diagram below.

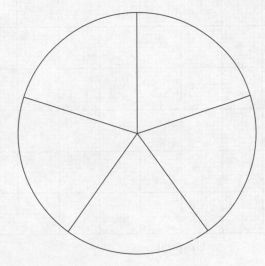

14. A recent study in Scotland looked at the average waiting times for people who went to the Accident and Emergency Unit at their local hospital. Some of the results are shown below.

Hospital	Average waiting time (minutes)
1	80
2	90
3	120
4	190
5	160

KU | PS

2

(a) Present these data as **a bar chart** using the grid below.

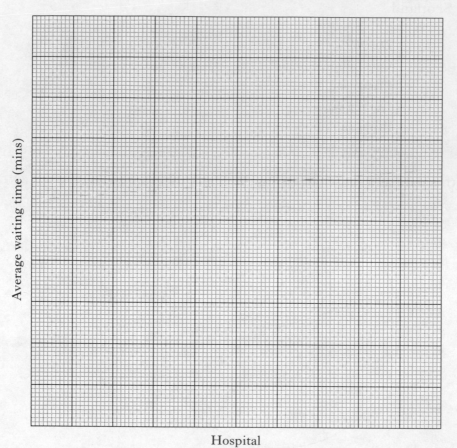

Hospital

(b) Calculate the average waiting time.

Space for calculation

_____ days

(c) How many more times longer would you have to wait in hospital 5 compared with hospital 1?

Space for calculation

_____

2

1

1

15. The diagram below shows a view of the internal structure of a leaf.

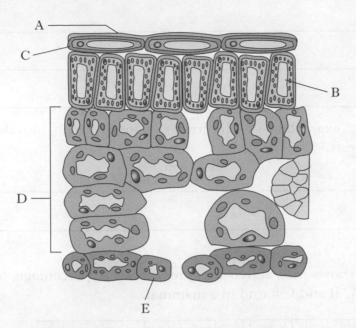

(a) Complete the following table.

Letter	Structure	Function/Description
A	Waxy cuticle	
B		Light can pass straight through this layer
C	Palisade mesophyll	
D	Spongy mesophyll	
E	Guard cell	

(b) State the substance which is found in the cell walls of xylem and gives it strength.

_____

(c) State **two** uses for the sugar made by photosynthesis.

1 _____

2 _____

KU: 3

KU: 1

PS: 2

	KU	PS

(d) (i) Explain what is meant by a limiting factor.

_____

_____

_____   1

(ii) Give **two** examples of limiting factors which would affect photosynthesis.

1 _____

2 _____   1

16. The graph below shows the activity, expressed as a percentage, of three different enzymes, A, B and C found in a mammal.

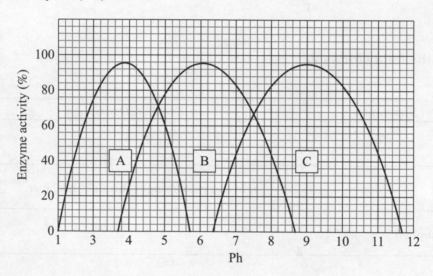

(a) Which enzyme is active over the smallest spread of pH? _____   1

(b) At which pH will **both** enzymes B and C work best? _____   1

(c) At what pH value is enzyme C at 60% of its maximum?

_____   1

(d) At pH 7, how many more times active is enzyme B compared with enzyme C at the same pH?

Space for calculation

_____   1

[End of Question Paper]

Exam B – Credit Level

Biology

Standard Grade: Credit

Practice Papers
For SQA Exams

Exam B
Credit Level

Fill in these boxes:

Name of centre

Town

Forename(s)

Surname

Try to answer all of the questions.

You have 1 hour, 30 minutes to complete this paper.

Write your answer in the spaces provided.

Leckie × Leckie

Scotland's leading educational publishers

1. (*a*) The diagram below shows some of the stages involved in making beer.

(i) In the space provided describe what is happening during **Stage 3**.

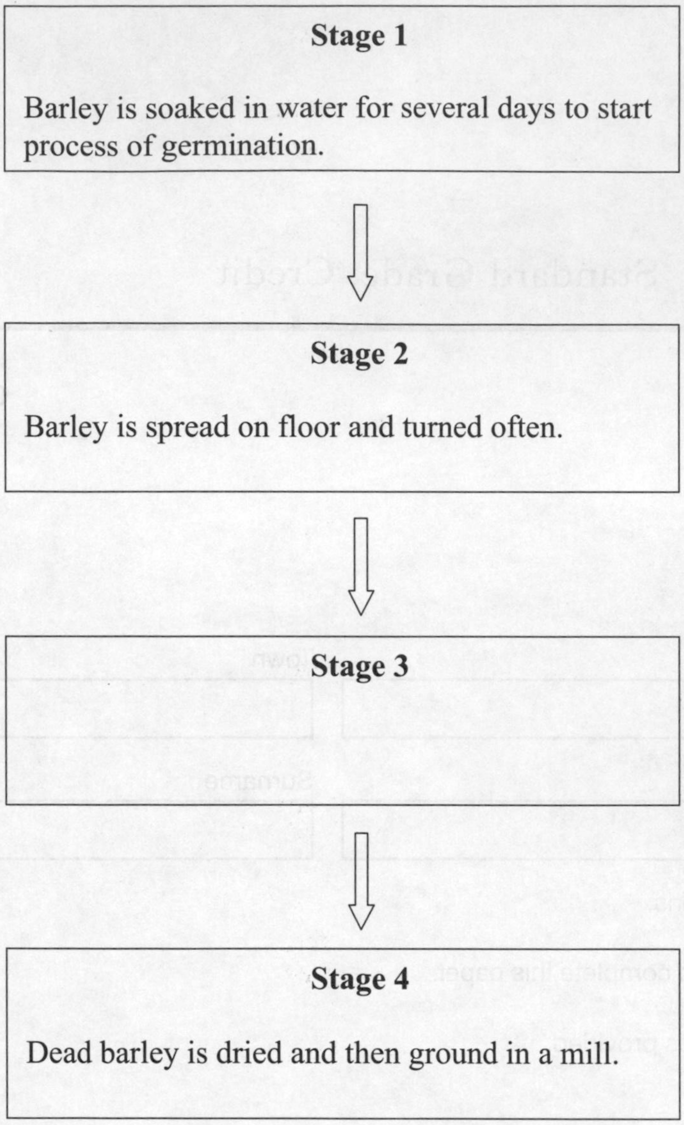

Stage 1
Barley is soaked in water for several days to start process of germination.

⬇

Stage 2
Barley is spread on floor and turned often.

⬇

Stage 3

⬇

Stage 4
Dead barley is dried and then ground in a mill.

(ii) As well as a food supply, state **two** conditions needed by yeast for growth.

1 _____

2 _____

KU · PS

1

2

(b) Describe how increasing temperature during summer might affect the percentage germination of seeds.

_____

_____

_____

(c) Explain how milk sours in terms of fermentation by bacteria.

_____

_____

_____

2. The data below shows the percentage composition of different nutrients in a plant food. The vitamin and mineral content was negligible.

Protein   40

Fat   25

Carbohydrate   15

Water   20

(a) Present this information as a **bar chart** on the grid below adding appropriate labels.

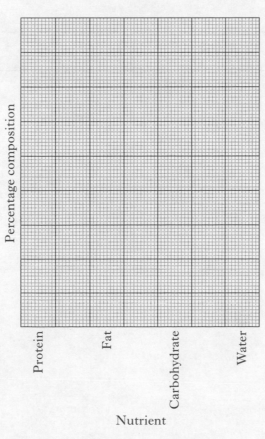

(b) Express the percentage compositions of protein, fat, carbohydrate and water as a simple whole number ratio.

_____ : _____ : _____ : _____

  protein   fat   carbohydrate   water

(c) What mass of fat would be found in 280g of this plant food?

Space for calculation

_____ g

3. State a technique which can be used to sample organisms. Give **two** possible sources of error when using this apparatus and state how each error can be reduced.

Technique   _____

1 Source of error   _____

_____

How to reduce this _____

_____

2 Source of error   _____

_____

How to reduce this _____

_____

KU	PS
	1
	1
2	

4. Read the following passage and then answer the questions that follow.

For thousands of years, humans have been selecting animals to produce better and more useful varieties. For example, animals with increased strength or which gave good yields of milk or which grew quickly. This was very much a "hit or miss" since nobody fully understood exactly why animals resemble their parents. It was not until relatively recently that the mechanism of inheritance began to be worked out by Mendel.

Mendel, working with the garden pea plant, discovered how inheritance worked by repeating experiments over many years to obtain a lot of consistent results. He used the pea plant because it had a number of easily observable differences or "traits" such as seed shape (wrinkled or round) or height (tall or dwarf) and so forth. By breeding experiments with these plants, he was able to suggest that traits which were inherited were controlled by "factors", which we now call genes, present in pairs. These factors exist in different forms, dominant and recessive. He always started his experiment with true-breeding plants which means they had the same two alleles, either both dominant or both recessive for the trait he was studying. Such individuals are known as homozygotes. By cross-pollinating two homozygotes of different appearance, such as tall and dwarf plants, he showed the first generation were always of the dominant tall appearance but carried the recessive allele without expressing it. These are called heterozygotes. There was never any dilution of the dominant appearance in the heterozygotes, so middle-sized plants were not found. However, when he cross-pollinated two of the first generation, he regularly obtained a ratio of 3 dominant to 1 recessive traits.

With a knowledge of basic Mendelian genetics, it is possible now to direct the process of selecting animals with desirable features. Providing the genetic makeup of the parent animals is known, animal breeders can predict the outcome of breeding experiments and so make these much more efficient. Such artificial selection also allows study of some of the processes involved in evolution.

(a) Give **two** improvements in animals which humans have artificially selected.

1 _____

2 _____   1

(b) Why was the early use of artificial selection a "hit or miss" approach?

_____

_____   1

KU	PS

	KU	PS

(c) What aspect of Mendel's experiments make his results very reliable?

_____

(d) Why was the pea plant such a suitable organism for Mendel's experiments?

_____

_____

(e) When starting a breeding experiment, what feature of the parent plants did Mendel ensure?

_____

_____

(f) How has a knowledge of basic Mendelian genetics helped animal breeders?

_____

_____

(g) In what other way mentioned does a study of artificial selection help scientists?

_____

5. (a) The grid below refers to the human digestive system.

A	B	C	D
Salivary glands	Stomach	Liver	Oesophagus
E	F	G	H
Gall bladder	Appendix	Large intestine	Rectum

Use the letters from the boxes to answer the questions which follow.

(i) Bile is produced in the [   ] and stored in the [   ].

(ii) Starch is converted to maltose by an enzyme produced by the [   ].

(iii) [   ] produces pepsin.

	KU	PS

(b) The diagram below shows one villus from a human small intestine.

Complete the table.

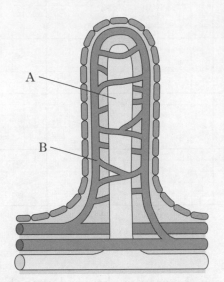

Structure	Name	Function
A	Lacteal	
B		

2

6. The following data shows the changes in dry mass of food eaten by two different species of fish in a laboratory experiment. The mass of food eaten was measured every week and the fish were fed exactly the same kind of food.

Species A	Time (weeks)				
	1	2	3	4	5
Dry mass of food eaten (mg)	8	20	12	5	10

Species B	Time (weeks)				
	1	2	3	4	5
Dry mass of food eaten (mg)	16	8	4	3	2

(*a*) Plot this data as two **line graphs** on the grid below.

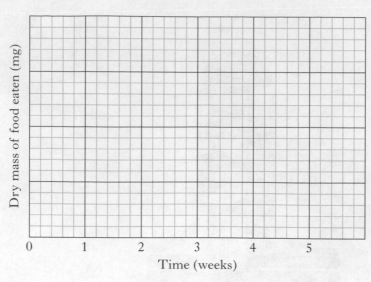

(*b*) Give **one** difference between the results obtained over the five week study.

_____

_____

(*c*) How much more food was eaten by species B compared with species A after one week?

_____ mg

(*d*) Predict what might happen to fish B if the experiment was allowed to continue.

_____

7. (*a*) Complete the table below which compares some aspects of reproduction and development in fish and mammals.

	Number of eggs produced	Length of time of parental care	Protection given to eggs/ young	Chances of young developing
Fish		Weeks	Almost none	
Mammal	Not many			High

KU | PS

3

1

1

1

2

(b) The placenta allows the exchange of substances between the mother and the developing baby. Write in four suitable examples of such substances in the boxes provided. The arrows indicate the direction of movement of the substances.

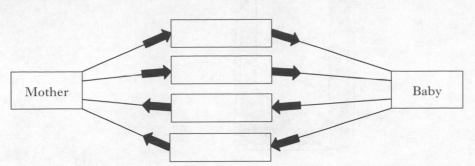

KU 2

8. The following bar chart shows the speeds of four different animals.

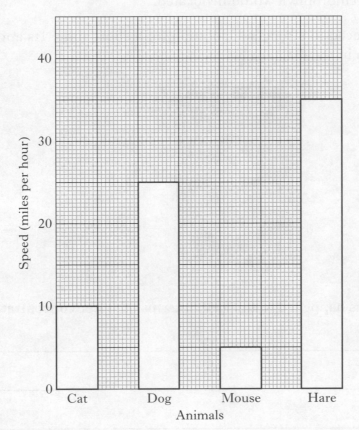

(a) What is difference in speed between the fastest and slowest animal?

Space for working

_____ mph

PS 1

(b) Express the speeds of the cat, dog, mouse and hare as a simple whole number ratio.

____ : ____ : ____ : ____
cat   dog   mouse   hare

PS 1

(c) Which animal can travel 3.5 times faster than a cat?

_____

PS 1

9. The diagram below shows a typical plant cell.

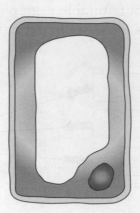

(a) Indicate by the letter "G" and a line where the genes controlling the manufacture of chlorophyll would be located.

(b) The cell was placed in strong salt solution for a short time. Its appearance now is as shown in the diagram below.

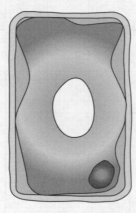

Explain what has happened to this cell in terms of water concentrations.

_____

_____

_____

10. The data below refers to the populations of a small mammal and its predator on an island. The data was collected in July of each year over a period of five years.

Year	1	2	3	4	5
Small mammal	700	650	600	150	500
Predator	14	21	15	10	5

(a) What is average number of the predator on this island each July?

Space for calculation

_____

(b) What is the relationship between the changing numbers of the small mammal and the numbers of the predator?

_____

_____

(c) A disease affected the small mammals on this island, killing many of them. The disease arrived in spring.

In which year did this happen? _____

(d) How could the results of this investigation be made more reliable?

_____

_____

KU	PS
	1
	1
	1
	1

11. (*a*) 1g each of carbohydrate, protein and fat were burned and the energy obtained recorded in kJ per gram in the following table.

Sample	Energy kJ/g
A	40
B	20
C	20

(i) Which of these samples was 1g of fat?

_____

(ii) How much energy would be supplied by 15g of sample B?

Space for calculation

_____ kJ/g

(*b*) Give **two** uses of the energy released from respiration.

1 _____

2 _____

(*c*) State the chemical elements found in carbohydrate.

_____

(*d*) What single term describes all the chemical reactions which go on inside a living cell?

_____

KU: 1, 1, 1, 1

PS: 1

12. The pie chart below shows the relative distribution of some diseases a group of 200 people had when they were young.

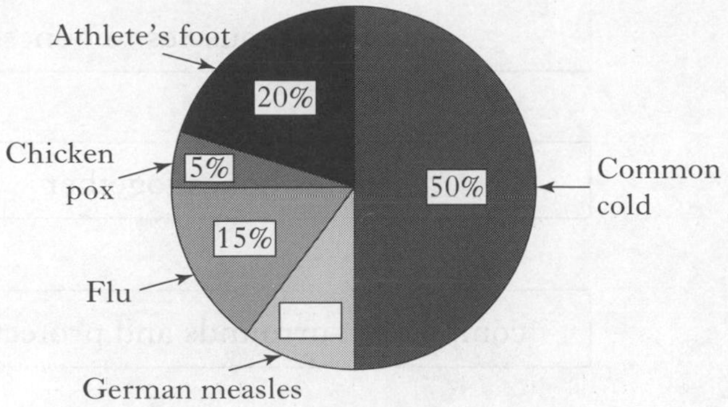

(a) In the box provided on the diagram, write the percentage of people who had german measles.

Space for working

(b) How many people had the common cold?

Space for working

(c) Express as a simple whole number ratio the percentages of people who had athlete's foot to flu.

_____ : _____
athlete's   flu
  foot

(d) If the sample size had been increased to 1000, how would this have improved the experimental procedure?

KU	PS
	1
	1
	1
	1

13. (*a*) Draw a line to connect the correct part of a synovial joint to its function.

Part	Function
capsule	connects muscles to bones
ligament	holds bones together
tendon	completely surrounds and protects joint

(*b*) Explain why muscles need to be in opposing pairs at joints.

_____

_____

_____

(*c*) Which part of a joint manufactures synovial fluid and give the function of this fluid.

Part of joint _____

Function _____

KU 2

KU 2

KU 1

14. The table below shows some data obtained when plants were grown in different nutrient solutions. For each nutrient solution, 20 plants were grown.

Nutrient solution	Dry mass (mg)		
	Leaves	Roots	Total
A		1400	3900
B	1814		2950
C		1200	3500
D	4000	3000	7000

(a) Complete the table by calculating the missing dry masses.

Space for calculation

(b) What is the average total dry mass of each plant grown in solution C?

Space for calculation

_____ mg

(c) What is the relationship between the dry mass of the leaves and the dry mass of the roots for each nutrient solution?

_____

(d) Which solution produced plants whose total dry mass was twice that of solution C?

Space for calculation

_____

KU | PS

2

1

1

1

15. (a) Explain the difference between continuous variation and discontinuous variation.

_____

_____

_____

2

(b) Complete the following table by putting a tick next to the correct type of change linked to the example given.

Example	Mutation	Selective breeding
Wheat for making bread has many sets of chromosomes and gives high yields		
Cauliflower has been produced from "ancient" cabbage plants over many generations		
Hundreds of years of careful matings has resulted in Friesian cows which give very creamy milk		
Disease-resistance in tomato plants		

2

(c) (i) What does the term _mutation_ mean? _____

_____

1

(ii) Give an example of a factor which can increase the rate of mutation in an organism.

_____

1

(d) What is meant by the term _monohybrid_ cross?

_____

_____

1

16. Ten pupils in a school, five boys and five girls and all the same age, were tested to see how much force, measured in newtons (N) they could apply to a scale using two different muscles. One muscle was in the arm and one in the leg.

The results of this investigation are shown below.

Boy	Force (N)	
	Arm muscle	Leg muscle
1	650	1600
2	475	1200
3	375	900
4	400	1100
5	500	1200
Average	480	

Girl	Force (N)	
	Arm muscle	Leg muscle
1	350	900
2	200	650
3	325	850
4	175	475
5	150	425
Average	240	660

(*a*) Complete the table by calculating average force of the leg muscle in the boys and writing this into the space provided.

Space for calculation

(*b*) Complete the **bar chart** of the averages for the results on the grid below.

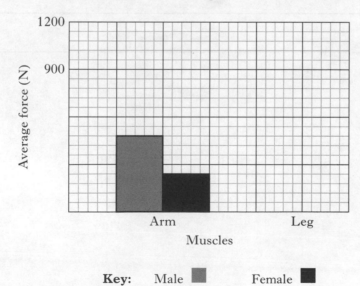

Key: Male ▨   Female ■

KU | PS

1

2

(c) How many times greater was the average force generated in the arm muscle of a boy compared with that of a female?

Space for calculation

_____

(d) State **one** conclusion about the two muscles in boys and girls.

_____

_____

(e) What is the combined force which boy 4 can exert using both the arm and leg muscles together?

Space for calculation

_____ N

17. (a) Describe **two** potential uses of plants or plant products.

1 _____

2 _____

(b) Explain how the loss of plant species might affect humans.

_____

_____

_____

[End of Question Paper]

KU	PS
	1
	1
	1
1	
1	
2	

Exam C – Credit Level

Biology

Standard Grade: Credit

Practice Papers
For SQA Exams

Exam C
Credit Level

Fill in these boxes:

Name of centre

Town

Forename(s)

Surname

Try to answer all of the questions.

You have 1 hour, 30 minutes to complete this paper.

Write your answer in the spaces provided.

1. (a) The table below shows some examples of animal responses to environmental stimuli.

Response to environmental stimuli
1 Woodlice avoid dry places
2 Maggots move towards decaying animals
3 Frogs stay near water

Choose **two** of these responses and explain how each helps the animal to survive. Use a number to indicate your choice in each case.

Response [ ]

Survival value _____

_____

Response [ ]

Survival value _____

_____

(b) The table below shows some examples of rhythmical behaviour in animals.

Rhythmical behaviour
1 Birds migrate from Scotland to Africa in autumn
2 Hedgehogs hibernate during winter
3 Cockroaches are more active at night

Choose **two** of these rhythmical behaviours and explain the significance of each to the animal's survival. Use a number to indicate your choice in each case.

Response [ ]

Survival value _____

_____

Response [ ]

Survival value _____

_____

KU PS

1

1

1

1

2. The data below shows some recordings of a pupil's breathing and pulse rates. During the recording session, the pupil rested for a few minutes, then did some exercise for a few minutes and then rested for a few minutes.

Time (mins)	Breathing rate (breaths/min)	Pulse rate (beats/min)
1	10	64
2	12	70
3	9	68
4	30	80
5	35	130
6	25	90
7	10	70

(a) Complete the plotting of the breathing and pulse rates as **line graphs** on the grid below.

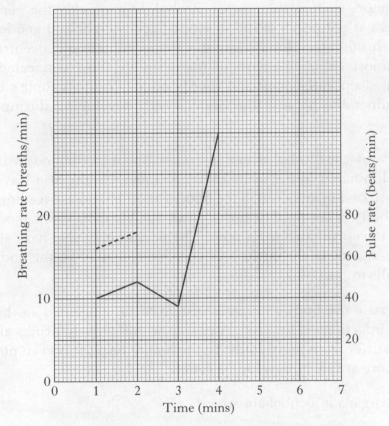

Key: Breathing rate ——— Pulse rate ------

(b) If the average volume of each breath during the first minute was 500 cm³, calculate the total volume of air the pupil breathed in.

Space for calculation

_____ cm³

(c) How could this experimental procedure be improved?

_____

KU | PS

3

1

1

(*d*) Between which of these time intervals did the pulse rate increase by 12 beats/min?

Underline the correct answer.

1 – 2 minutes

3 – 4 minutes

4 – 5 minutes

5 – 6 minutes

3. Read the following passage and then answer the questions that follow.

After swallowing alcohol, it travels to the stomach and small intestine and is absorbed from both into the bloodstream. The rate of this absorption is affected by a number of factors such as the type of alcohol, how quickly the person is drinking, if the person's stomach is full or empty, age, weight and gender. As the alcohol moves through the bloodstream it is found in highest concentrations in the brain. The most vulnerable parts of the brain are those connected with memory, attention, sleep, judgement and coordination. Young people's brains are particularly vulnerable because the brain is still developing during their teenage years.

Approximately 95% of the alcohol drunk is broken down in the body with the remainder being eliminated in the urine by the kidneys. Some also leaves the body in sweat, saliva, breath and even in the milk of mothers breast-feeding their babies. Since the body takes time to break down the alcohol, the person may appear drunk and uncoordinated for a time. The person may also have slurred speech and double-vision. If the intake was very large, it may be the body just cannot cope and falls into a coma.

Alcohol can also cause the body's immune system to be weakened so that the ability to fight off infections is lowered. A person who regularly drinks alcohol may, for example, suffer from more colds. In the longer term, it is even possible a person may be more at risk of developing cancer.

(*a*) By which **two** organs is alcohol absorbed?

1 _____

2 _____

(*b*) Give **two** factors which influence how quickly alcohol is absorbed.

1 _____

2 _____

(*c*) Why are young people particularly at risk from drinking too much alcohol?

_____

KU	PS
	1
	1
	1
	1
	1

	KU	PS

(d) Why might it be possible to detect alcohol in the sweat of a person who has been drinking?

_____

_____ **1**

(e) Give **two** reasons why it is not a good idea to operate dangerous machinery after drinking alcohol.

1 _____

_____

2 _____

_____ **2**

(f) Why are regular drinkers of alcohol more liable to develop colds?

_____

_____ **1**

(g) Give **one** long-term possible effect of drinking alcohol.

_____ **1**

4. (a) The grid below refers to the role of anti-diuretic hormone (ADH) in the control of water balance.

A	**B**	**C**	**D**
Blood too dilute	Large volume of water drunk	Sweating	Less ADH produced
E	**F**	**G**	**H**
More ADH produced	Small volume of urine produced	Large volume of urine produced	Blood too concentrated

Use the letters to answer the questions which follow.

(i) Complete the flowchart showing some of the changes in blood water levels.

Normal blood concentration ⇒ B ⇒ ☐ ⇒ ☐ ⇒ ☐ ⇒ Normal blood concentration **1**

(ii) What would be the immediate effect of eating very salty nuts? ☐ **1**

	KU	PS

(b) Give **one** advantage and one disadvantage of kidney transplantation for people whose kidneys have failed.

Advantage _____

_____

Disadvantage _____

_____

2

5. The following data shows the changes in mass of potato cores after being kept for twelve hours in four different concentrations of sugar.

Concentration of sugar (g/l)	Mass of potato cores (g)	
	At start of experiment	After twelve hours
0	3.2	4.0
50	3.5	4.2
100	3.5	2.8
150	2.8	2.1

(a) Complete the following table showing the percentage increases/decreases for each concentration of sugar.

Concentration of sugar (g/l)	Percentage change in mass
0	+25
50	
100	−20
150	

1

Space for calculation

(b) Plot the percentage change as a **line graph** on the grid below.

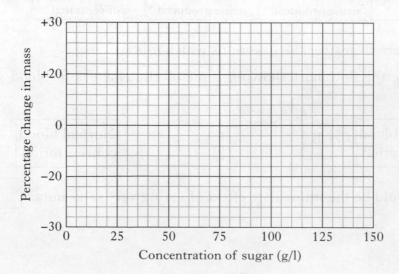

2

6. (*a*) John is exercising by weight-lifting but eventually finds he cannot do any more without a break.

In terms of anaerobic respiration, explain why John has to stop.

_____

_____

(*b*) State **two** ways training can improve body function.

1 _____

2 _____

(*c*) The graphs below show the output of the heart in litres/minute (l/min) with changing heartbeat measured in beats/minute (bpm) for a trained athlete and an untrained athlete.

(i) What is the heart output of each athlete at 80 bpm?

Trained athlete _____ l/min

Untrained athlete _____ l/min

(ii) What is the increase in heart output for the trained athlete when the heart rate increases from 90 to 100 bpm?

_____ l/min

(iii) What relationship is there between the heartbeat of a trained athlete and the heartbeat of an untrained athlete?

_____

_____

	KU	PS
	2	
	1	
		1
		1
		1

7. (a) The following statements refer to batch or continuous processing. By means of lines, connect the correct statements to each process.

KU	PS

Raw materials are all placed into a fermentor

BATCH PROCESSING

Reaction vessel is closed and left until process is complete

Whole cells are immobilised

CONTINUOUS PROCESSING

Product flows from the process

Enzyme can be reused

More cost-effective

2

(b) The diagram below shows an experiment into respiration by yeast cells in the absence of oxygen.

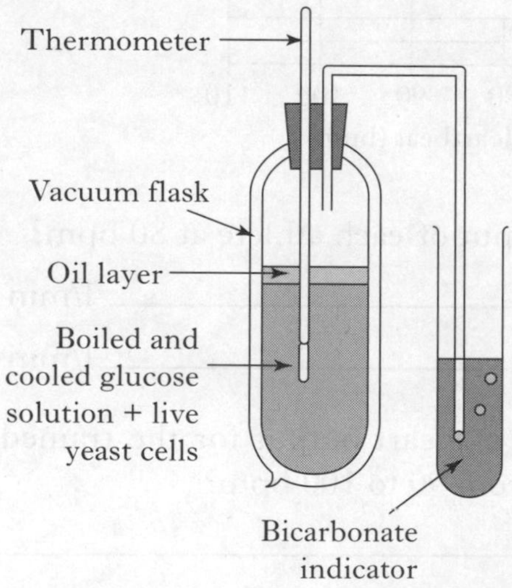

Thermometer

Vacuum flask

Oil layer

Boiled and cooled glucose solution + live yeast cells

Bicarbonate indicator

(i) Give **one** reason why the glucose solution was boiled before use.

_____

1

(ii) Why was the experiment carried out in a vacuum flask?

_____

_____

1

(iii) What is the function of the oil layer?

_____

1

(iv) What is the reason for including bicarbonate indicator in this experiment?

_____

(v) How would a control experiment differ from that shown?

_____

8. The data below shows the number of living cells of a micro-organism over a period of 7 hours.

Time (hours)	0	1	2	3	4	5	6	7
Number of living cells	20	40	70	120	170	240	300	350

(a) Plot these data as a **line graph** on the grid below.

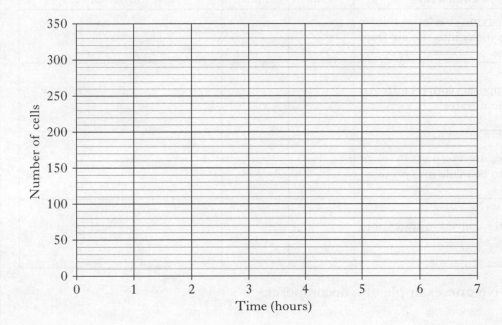

(b) Between which of these hours was the growth most rapid? Tick the box.

0 – 1 ☐

2 – 3 ☐

5 – 6 ☐

6 – 7 ☐

KU	PS
1	1
	2
	1

(c) Assuming nothing was added to the micro-organisms, predict what would happen to the number of living cells after 48 hours.

Explain your answer.

Prediction _____

Explanation _____

_____

9. (a) Some of the following statements about the nitrogen cycle are made false by incorrect terms which have been underlined.

If the statement is correct, tick the box marked **true** but if the statement is incorrect, tick the box marked **false** and write in the correct term(s) into the box provided.

Statement	True	False	Correction
<u>Denitrifying</u> bacteria release nitrogen gas into the air from nitrates			
<u>Decomposers</u> convert nitrogen gas into nitrates			
<u>Nitrites</u> are taken up by plant roots to be made into protein			
Bacteria convert <u>nitrates into nitrites</u>			

(b) (i) Give **one** example of a decomposer.

_____

(ii) State **two** environmental conditions which are required before decomposition can take place.

1 _____ 2 _____

(c) What are the structures called, present in legumes such as peas and beans, which contain nitrogen-fixing bacteria?

_____

KU	PS
	1
3	
1	
1	
1	

10. A group of people completed questionnaires about their eyesight and the following results were obtained.

Description of eyesight	Number of people
Short-sight	80
Long-sight	40
Other problems	20
Normal sight	60

(a) Use the data to complete the pie-chart below.

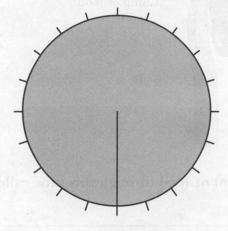

(b) How many people in total were sampled? _____

(c) What was the percentage of people who had long-sight? _____ %

(d) Express the number of people who had short-sight, long-sight, other problems and normal sight as a simple whole number ratio.

      _____ : _____ : _____ : _____

        short   long   other   normal

        sight   sight   problems   sight

(e) Which group had three times as many as those with 'other problems'?

_____

(f) How could this study be made more reliable?

_____

KU	PS
	1
	1
	1
	1
	1
	1

11. (a) The diagram below shows part of the tube which carries food from the mouth to the stomach.

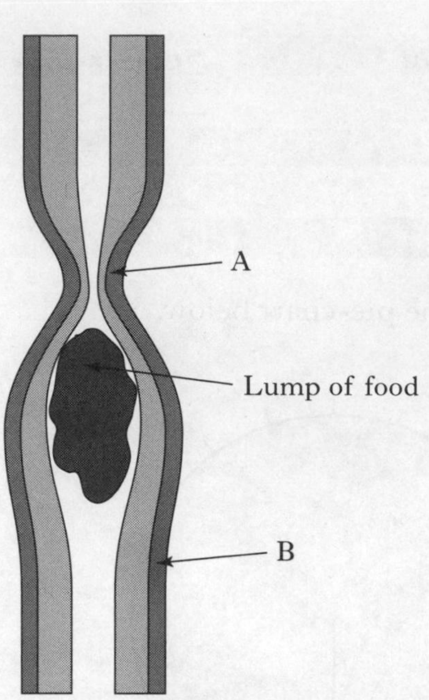

(i) What is this movement of food through this tube called?

_____

(ii) Complete the following table.

State of muscle	Result
A Contracted	
B	Allows lump of food to move easily

(iii) On the diagram indicate, by means of arrows, the direction of movement of the lump of food.

(b) How do the contractions in the muscles of the stomach wall help in the chemical breakdown of food?

_____

_____

_____

(c) Where is bile manufactured?

_____

KU | PS

1

1

1

2

1

12. A dog and cat home collected data for the intake of these animals over a period of three years and the results are shown below. The months are indicated by the first letter starting with "J" for January.

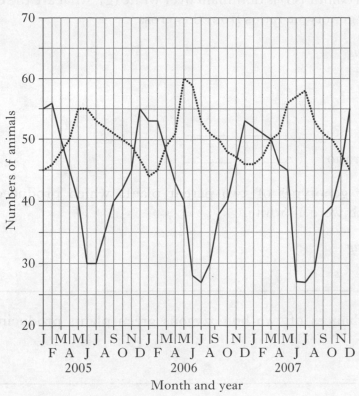

Key: Dog intake ———— Cat intake ·············

(*a*) In which months during 2005 were 40 dogs taken in? _____ | 1

(*b*) How many cats were taken in during the month of May 2006? _____ | 1

(*c*) When was the lowest intake of dogs? _____ | 1

(*d*) At what season of the year (spring, summer, autumn or winter) is the intake of each animal highest?

Cats _____ Dogs _____ | 2

13. A heterozygous green flowered plant was crossed with a homozygous white flowered plant.

(a) Assuming green colour (G) is dominant over white (g), what are the expected phenotypic and genotypic ratios from this cross?

Space for working

   (i)   Phenotypic ratio   _____ : _____

   (ii) Genotypic ratio   _____ : _____

(b) Give the likely genotypes of two parent plants which produced an F_1 generation which were all white.

Space for working

_____

(c) What are the chances of two homozygous green plants producing white plants?

_____

(d) If two heterozygous plants were crossed and produced 76 plants, how many of these 76 plants would you expect to be green?

Space for working

_____

(e) Give **two** reasons why it is possible that the expected ratios may be different from the observed.

1 _____

2 _____

KU	PS
2	
	1
	1
	1
2	

14. (*a*) The following stages are used in producing genetically engineered insulin.

A	Insulin is extracted and purified
B	Bacteria are grown
C	Plasmid and inserted human gene put into bacterial cell
D	Desired human gene is inserted into the plasmid
E	Bacterial plasmid is cut open
F	Desired human gene is identified and cut out of the chromosome

Using the letters, arrange this in the correct sequence.

F ⇨ ☐ ⇨ ☐ ⇨ ☐ ⇨ ☐ ⇨ ☐

(*b*) (i) Why is there an ever increasing need for insulin produced by genetic engineering?

_____

_____

(ii) Give **two** advantages in producing such a product by genetic engineering.

1 _____

2 _____

(*c*) State **two** advantages of using biological detergents containing enzymes which can work at relatively low temperatures such as 40°C.

1 _____

2 _____

KU PS

1

1

2

2

15. The diagram below shows tissue involved in the transport of food.

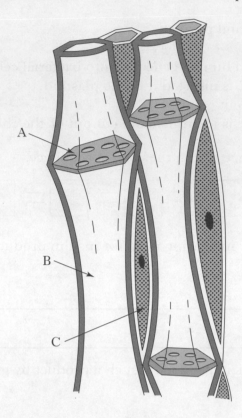

(a) Which letters indicate the companion cell and sieve plate?

Companion cell _____

Sieve plate _____

(b) What are the cells which control stomatal opening and closing called?

_____

[End of Question Paper]

KU	PS

1

1

Exam D – Credit Level

Biology

Standard Grade: Credit

Practice Papers
For SQA Exams

Exam D
Credit Level

Fill in these boxes:

Name of centre

Town

Forename(s)

Surname

Try to answer all of the questions.

You have 1 hour, 30 minutes to complete this paper.

Write your answer in the spaces provided.

Scotland's leading educational publishers

1. (a) Complete the following table which describes the action of the main structures involved in breathing.

Structures	Inhalation	Exhalation
Intercostal muscles		Relaxed
Ribcage	Moves up and out	
Lungs	Inflate	
Diaphragm		Relaxes and moves up

(b) Give **two** features of the lungs which make them efficient for gas exchange.

1 _____

2 _____

(c) The diagram below shows some of the cells which line the trachea.

B [   ]

[   ] A

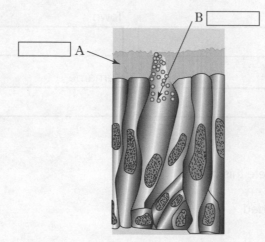

(i) Add suitable names for structures labelled A and the sticky material labelled B.

(ii) Explain the role of A and B in the trachea.

_____

_____

_____

KU: 2, 1, 1, 2
PS:

(d) Where is haemoglobin found and what is its function?

Where found _____

Function _____

_____

2. Read the following passage and then answer the questions that follow.

We spend anything up to one-third of our lives asleep. Trying to explain what sleep is and why it is so important continues to be a puzzle to sleep scientists. All animals sleep, even aquatic mammals such as dolphins which move all the time. They can do this by closing one eye and allowing one half of its brain to go to sleep.

What is well-known is that losing sleep can have serious consequences. Indeed, animal studies have shown that depriving rats and insects of sleep over a period of around two weeks will result in their death. In humans, short-term loss of sleep will result in stress – emotional, and physical as well as intellectual. Yet nobody fully understands how sleep restores the brain every night.

Sleep may also be important for organising new memories and allowing the brain to let go of random unimportant experiences. This leaves room for new learning to take place the next day. Sleep may also be a way of conserving energy since our feelings of hunger get switched off and so we don't need as much food. People who are deprived of sleep have been shown to increase their daily calorie intake.

Another theory is that humans are more vulnerable at night which means that in our evolutionary past, being asleep meant being less likely to be eaten by a predator. Our most important sense is sight which does not work well at night.

One obvious reason why sleep is so difficult to study is that scientists can't know what is going on in somebody's head when they are asleep. It is also ethically difficult to study sleep because it means keeping an animal or human awake, for example by using cold water thrown at the subjects.

(a) Approximately how much of our lives do we spend sleeping?

_____

KU: 2

PS: 1

	KU	PS

(b) How do aquatic mammals sleep yet are still able to move all the time?

_____

_____   **1**

(c) What is the effect of sleep deprivation in excess of two weeks on an animal such as a rat?

_____   **1**

(d) How does sleep possibly allow new learning to take place?

_____   **1**

(e) Give **two** ways in which sleep helps conserve energy.

1 _____

2 _____   **1**

(f) In what **two** ways mentioned are humans more vulnerable at night?

1 _____

2 _____   **1**

(g) Give **one** reason why scientists find studying sleep so difficult.

_____   **1**

(h) What ethical issue is mentioned regarding the study of sleep?

_____   **1**

	KU	PS

3. (a) The grid below refers to some aspects of pollution.

A	B	C	D
Very long lasting	Sewage	Pesticides	Lead

E	F	G	H
Sulphur dioxide	CFCs from aerosols	Global warming	Oil

Use the letters to match with the descriptions below.

(i)   Major cause of acid raid

(ii) Source of domestic pollution

(iii) Affects the ozone layer

(iv) Waste from nuclear power station

2

(b) What is meant by the term "crop rotation"?

_____

_____

1

(c) The following chemicals are used in farming.

A   Artificial fertilisers

B   Fungicides

C   Herbicides

D   Insecticides

Which chemicals would be used for each of the following? **Use the letter**.

Replace nitrates lost in soil _____

Kill weeds _____

1

	KU	PS

	KU	PS

4. (a) The diagram below shows a red blood cell in different solutions.

A

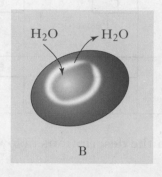
B

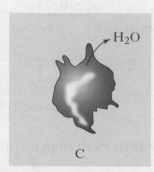
C

Which solution has a high water concentration? Explain your answer.

Solution _____

Explanation _____

_____

_____

(b) What is the importance of diffusion to cells?

_____

_____

_____

2

1

5. The graph below shows the population of a country in two different years, measured in millions.

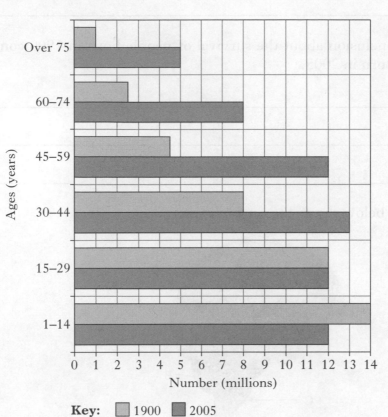

Key: ☐ 1900   ■ 2005

(a) What was the total population in 2005?

Space for working

_____ millions

(b) What percentage of the population in 2005 was 75 or older?

Space for working

_____ %

(c) In 1900 it was found that 8 million of the population were male aged 14 years or younger.

In the same year, how many females aged 14 years or younger were there?

Space for calculation

_____ millions

KU | PS

1

1

1

1

(d) Which age range showed the same number for both 1900 and 2005?

_____ years

(e) Draw **one** conclusion about the survival of people born in 1900 compared with people born in 2005.

_____

_____

_____

6. (a) The diagram below shows the human brain.

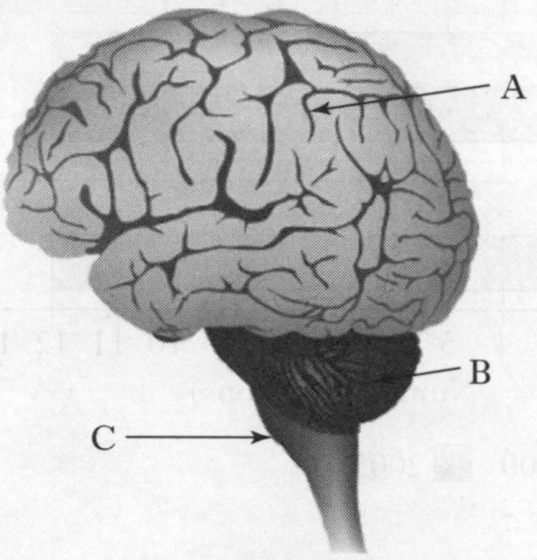

Complete the following table.

Part	Name	Function(s)
A		Memory, thought, intelligence, personality
B	Cerebellum	
C		

(b) Explain how the arrangement of the semi-circular canals in the ear is related to their function.

_____

_____

_____

KU	PS
	1
	1
2	
2	

(c) How does binocular vision help humans judge distance?

_____

_____

_____

7. (a) The diagram below shows some different kinds of fruits.

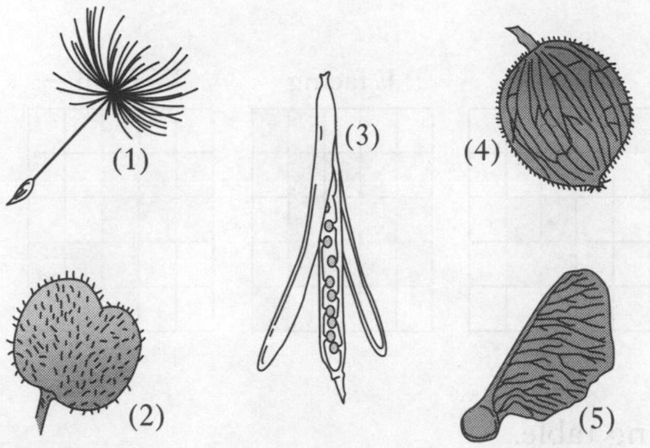

Complete the table to show how these fruits are dispersed. **Use the numbers**.

Animal internal	Animal external	Explosive	Wind

(b) Match the following descriptions of flowers with the method of pollination by drawing lines to connect each description to the correct method.

| Small quantities of pollen produced |

| Anthers large and loose |

| Brightly coloured petals |

| No nectaries are present |

| Stigmas large and feathery |

WIND POLLINATION

INSECT POLLINATION

KU | PS

2

1

2

(c) Give **one** advantage of grafting plants.

KU	PS
1	

8. A tree was sampled for the presence of the simple plant *pleurococcus* by using small quadrats. Each sample was taken at four different compass points, north [N], south [S], east [E] and west [W].

The results are shown below, the shading indicating the presence of the plant.

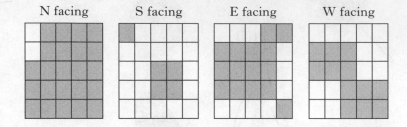

N facing   S facing   E facing   W facing

(a) Complete the following table.

Space for calculation

Facing	% cover
North	92
South	
East	
West	48

(b) Another tree gave a 96% for the north-facing sample. Indicate, by shading the appropriate number of squares, a possible result on the blank quadrat below.

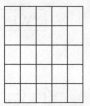

(c) Suggest a relationship between the compass points and the presence of the *pleurococcus*.

_____

_____

_____

1

(d) How could this experimental procedure be improved?

_____

_____

1

9. The diagram below shows some aspects of the carbon cycle.

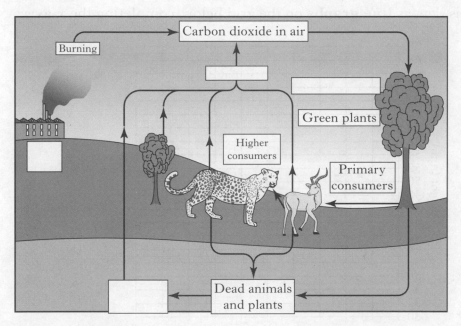

(a) Insert the following terms into the correct boxes.

Fossil fuels

Respiration

Decay by bacteria

Photosynthesis

2

KU | PS

(b) What would happen if decay by bacteria did not take place?

_____

_____

1

10. Two seedlings, A and B, were grown. Both were given identical volumes of water but seedling A had some dissolved chemical fertilizer added.

The heights of each seedling over a period of ten days are shown below.

Time (days)	1	2	3	4	5	6	7	8	9	10
Height seedling A (mm)	50	52	56	58	60	62	62	64	66	70
Height seedling B (mm)	40	42	44	45	48	50	52	52	54	60

(a) Draw this data as **line graphs** on the grid below, completing the x-axis.

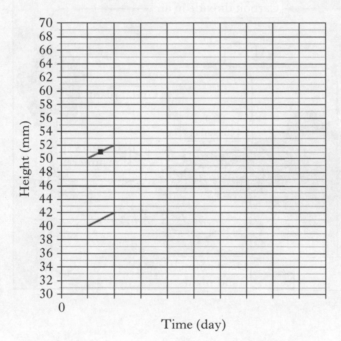

Key: Seedling A ▬■▬   Seedling B ▬

3

(b) (i)   What was the percentage increase of each seedling over the ten days?

Seedling A _____ %

Seedling B _____ %

2

(ii) Suggest a reason for the difference.

_____

_____

1

(c) State **one** factor, apart from the volume of water added, concerning the environmental conditions which should be kept constant.

_____

(d) How could the reliability of this experiment be improved?

_____

_____

11. (a) The diagram below shows part of a food web.

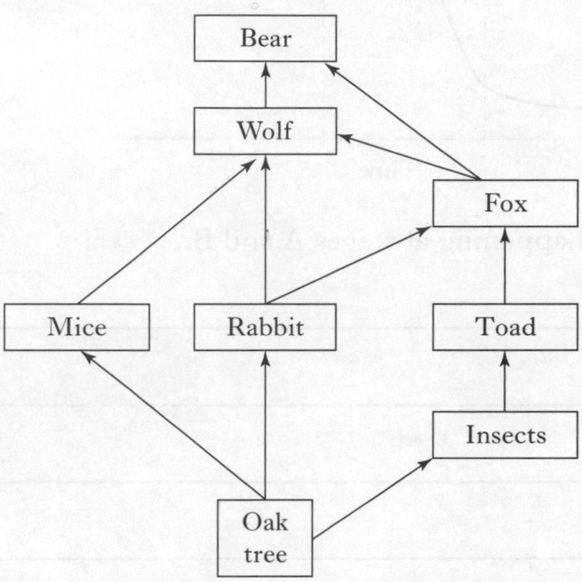

(a) What would be the short-term effect on the wolf population if the bears were removed from this food web?

_____

(b) Explain what is meant by a pyramid of numbers.

_____

_____

(c) How can plants such as clover obtain nitrates in soil which is low in nitrates?

_____

_____

KU	PS
	1
	1
1	
1	
2	

(c) The diagram below shows three stages in the growth of a population.

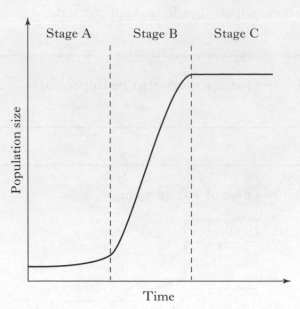

(i) Explain what is happening at stages A and B.

Stage A _____

_____

Stage B _____

_____

(ii) What is the relationship between the birth rate and death rate during stage C?

_____

12. The graph below shows the food content of the milk of two mammals.

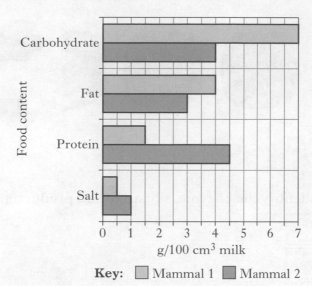

Key: ☐ Mammal 1   ■ Mammal 2

(a) How does the salt content of the milk of mammal 1 compare with mammal 2?

_____

(b) Suggest a possible reason why the milk of mammal 2 has a higher protein content than the milk of mammal 1.

_____

_____

(c) Express as a simple whole number ratio the carbohydrate and fat content of mammal 2.

_____ : _____

KU	PS
	1
	1
	1

13. A possible mechanism for the action of enzymes is shown below.

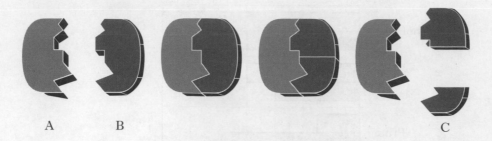

A   B   C

(a) **Using the letters** identify the enzyme, substrate and product(s).

Enzyme ☐

Substrate ☐

Product(s) ☐

1

(b) The enzyme pepsin is said to be "specific" for protein. What does this mean?

_____

1

(c) This enzyme, which comes from the human body, has an optimum activity at 37°C.

Explain what is meant by an enzyme's "optimum" temperature.

_____

_____

1

KU	PS

14. Twenty pupils, ten male and ten female, aged fifteen years, had their heights measured in cms. The results are shown in the table below.

Male heights (cms)	180	178	175	190	185	180	175	178	179	180
Female heights (cms)	154	161	163	157	161	164	158	156	162	164

(a) What is the range of heights for the male and female pupils?

Male pupils _____ cm to _____ cm

Female pupils _____ cm to _____ cm

(b) Complete the x-axis and plot these averages as a **bar chart** on the grid below.

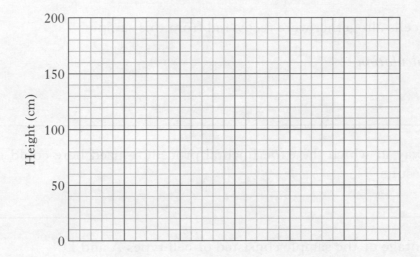

(c) State **one** way of ensuring the experimental procedure is reliable.

_____

KU	PS
	2
	2
	1

15. A small sample of material from a plant yielded the following four different cell types, A, B, C and D in the numbers shown in the table below and plotted on the pie-chart.

Cell type	Numbers
A	86
B	20
C	60
D	34

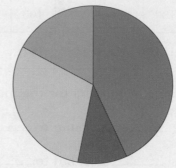

(a) On the pie-chart write the **letters** of the four different cell types into the correct segments.

(b) How many cells altogether were recovered from this sample?

Space for calculation

_____

(c) State **one** way in which the experimental procedure used here could give unreliable results.

_____

(d) What percentage of the sample consisted of cell types A and D?

Space for calculation

Cell type A _____ %   Cell type B _____ %

(e) Express as a simple whole number ratio the different cell types.

$$\underline{\hspace{1cm}} : \underline{\hspace{1cm}} : \underline{\hspace{1cm}} : \underline{\hspace{1cm}}$$
$$\;\;A\qquad B\qquad C\qquad D$$

16. Explain the importance of internal fertilisation to land-dwelling animals.

_____

_____

_____

_____

[End of Question Paper]

KU   PS

1

1

1

2

1

2

Worked Answers A – D

1. (*a*)

> This question tests your ability to read a paired statement key which is a problem solving ability. An important point about the construction of the statements is that they should be "opposites" of each other. For example, in a key dealing with birds, you might say "wings present" but then the other linked statement should read "wings absent" not, for example, "feathers present". This is a common error and one you should take care to avoid. Another source of error is going beyond the key. For example, in a key dealing with vertebrates if you were asked to describe a rattlesnake and said "forked tongue present" you would need to be sure that this was in the information given. If not, even though it is correct, you would get no award since the question is testing your ability to use the key and information given properly.

(i) Using this information, complete the boxes in the paired statement key below

4	Legs present	go to 2
	Legs absent	go to 4

2	Lives in sea	*crayfish*
	Does not live in sea	go to 3

3	Six legs present	*ant*
	More than six legs present	*scorpion*

4	Segments present	*earthworm*
	Segments absent	*ragworm*

[3] PS

> **HINT**   5 correct = 3 marks/3 or 4 correct = 2 marks/2 correct = 1 mark

(ii) Similarity   both have legs/both have a body which is hard and segmented

Difference   scorpion lives on land, crayfish lives in sea

[1] PS

> **HINT**   Both answers are needed for 1 mark

(*b*) indicator species

[1] KU

> **HINT**   Notice that the presence OR absence of a particular organism can indicate the state of a part of the environment

2. (*a*) (i)

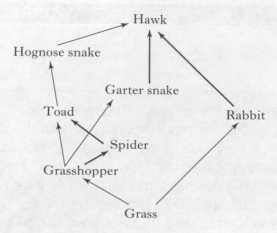

[2] PS

> **HINT**
>
> Remember when completing a food web, the arrowed lines indicate the flow of energy so be careful with the direction. The partially completed web makes this easy. You can see, for example, the hawk eats the hognose snake which in turn eats toads.
>
> Notice that one arrowed line (between grasshopper and garter snake) crosses another (between spider and toad). This is perfectly in order. Four correct arrowed lines are needed to obtain all 2 marks and usually 3 or 2 correct arrowed lines would get 1 mark with only 1 correct arrowed line obtaining no score. There are NO half marks available in Standard Grade Biology!

(ii) The population of hognose snakes would

increase

<u>decrease</u>

stay the same

In this food web, the hognose snake has only one food source, the toads, and if these are removed, the snake would have nothing to eat

[1] KU

> **HINT**
>
> Both answers are needed for 1 mark.

TOP EXAM TIP

Throughout the paper you will see some instructional words emphasised. For example, "**one**" or "**underline**" to give you a little guidance as to how to answer particular questions. Make sure you always follow this guidance.

(*b*) the number of organisms decreases while the size of them increases

[1] KU

> **HINT**
>
> When you are asked, as in this question, to give the relationship between two variables, make sure always to state the effect of any change in one on the other. Do not, for example, simply say "the number of organisms decreases."

3.

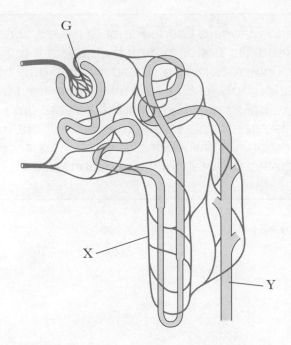

(a) (i) nephron

[1] KU

(ii) [see diagram above]

[1] KU

> HINT
>
> It is conventional when labelling a biological diagram, to use lines without arrow heads as is done in the diagram above. You would not lose marks if you did but try to observe the correct convention, making sure the line ends clearly on the structure you intend.

(iii) X reabsorption/useful materials are taken back into the blood

Y collects urine from other nephrons/transports urine to bladder

[2] KU

> HINT
>
> Pay attention always to what the question is asking. Here you are not asked to name structures X and Y but only to give their respective functions.
>
> 1 mark for each correct answer

(b) by the breakdown of excess amino acids

[1] KU

> HINT
>
> It is important to use the word "excess" in this answer.

4.

It is usual now for each Standard Grade Paper to have a short scientific passage for you to read. Do not be put off by the fact it may deal with something you have not met before. Basically, the questions which follow the passage are simply testing your ability to extract the relevant answers from the text, not to recall anything. Take your time and make sure what you write is appropriate to the question without missing out anything. Where appropriate, include the units such as seconds / °C / % and so forth. Notice that the order of the questions usually follows the sequence of the passage.

(*a*) not more than a day or so

[1] PS

(*b*) 65% of its body weight

[1] PS

(*c*) they get their water from their food

[1] PS

(*d*) 1   support

2   steering

[1] PS

HINT   Both answers needed for 1 mark

(*e*) it comes out for food only at night when it is cooler and water loss is reduced

[1] PS

(*f*) it has only a short range of movement

[1] PS

(*g*) seeds and other dry plant material

[1] PS

5. (*a*) D

this is the next most effective antibiotic

[1] PS

HINT   Both answers are needed for 1 mark

(*b*) C

[1] PS

(c) bacteria are sensitive to some antibiotics and not others/bacteria may develop resistant forms

[1] KU

(d) by repeating the procedure/using more antibiotics

[1] PS

6.

It is true to say that genetics causes problems for many students yet, with a lot of practice, the questions are usually not that challenging. They usually follow a set pattern, such as shown below, with some variations. Care must be taken to ensure letters used to represent different forms of a gene, called alleles, are not confused with each other. Here it would be easy to mix up "O" with "o." Do not cut corners with genetics questions and always use the Punnett square to make sure you have worked through the questions properly.

(a) (i)

	Genotype	Phenotype
Parents	OO and oo	orange and cream
F_1	all Oo	orange

[2] KU

HINT   1 mark for each correct line

(ii) Genotypic ratio   1 OO : 2 Oo : 1 oo
      Phenotypic ratio   3 orange : 1 cream

[2] KU

HINT   1 mark for each correct ratio

When you are asked, as here, to "predict" what will happen, this is asking you to make a best scientific guess based on the data you have. Sometimes you may be asked why your prediction does not match what actually happened.

Do not get confused between the "phenotype" and "genotype" of an organism. Phenotype is an expression of the genes, here the colour of the fruit. Genotype is the combination of the alleles, such as OO or oo or Oo here.

(iii) ¼ of 500 = 125

[1] PS

HINT   A ratio is always expressed with the numbers separated by a colon such as 3 : 2 or 4 : 7. Try and give the ratio in simple whole numbers, such as 2 : 1 rather than 4 : 2.

(b) alleles

[1] KU

7.

"Data" is a general term for figures related to some experiment or measurements taken. It may be presented in different ways, usually indicated in the question by bold text. A bar chart needs to drawn with a ruler, each axis must be properly labelled with units included where appropriate. Each bar should be the same width and separated from the next bar by equal spacing. Make sure you use more than half the grid. Take time over this type of question because students are often careless with these points. If you do make a mess of the grid, there will be spare grids included at the end of the paper. Usually, one mark is given for the correct labelling of the axes and one mark for the accurate plotting of the data.

(*a*)

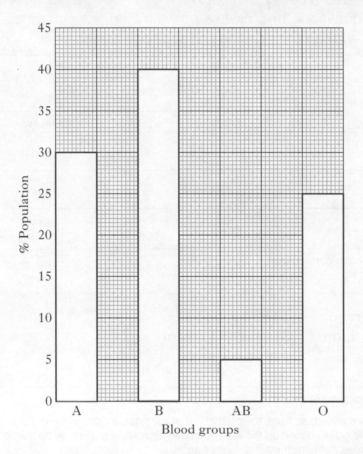

[2] PS

HINT   1 mark for correct completion of each axis and 1 mark for correct drawing of bars

(b) 60

[1] PS

> **HINT** — From the data in the table we know that 5% of the human population has blood group AB. To calculate the predicted number in the town, it would be 5% of the town's population which is 1200. Therefore, $[5 \times 1200] \div 100 = 60$.

(c) size of the sample is too small

[1] PS

> **HINT** — In any scientific investigation, the larger the sample size, the more reliable the results will be.

8. (a) petals
usually brightly coloured which attracts insects

[1] KU

> **HINT** — You must use the diagram to answer this first question. While nectaries, for example, attract insects, you cannot see these on the diagram so do not use these as an example. Similarly, don't give scent as a "structure" to attract insects.
>
> Both answers are required for the mark.

(b) insect-pollinated flowers have small petals/dull petals/stamens inside/stigma small/stigma inside/pollen large/pollen sticky/nectaries present

wind-pollinated flowers have large petals/brightly coloured petals/stamens hanging outside/stigma large/stigma hanging outside/pollen small/pollen light/nectaries absent

[2] KU

> **HINT** — In this question, make sure to include contrasting pairs of features. Thus, if you say "stigma small" for insect-pollinated flowers, make sure you contrast this with "stigma large" for wind-pollinated flowers.
>
> 1 mark for each correct pair of differences.

(c) clone

[1] KU

(d) transport of water/support

[1] KU

9.

The advice given earlier for drawing bar charts applies here to line graphs equally. Make sure you indicate the points plotted clearly, join them with a straight line. Do not make a smooth curve or best fit graph. Do not plot points for which you have no data. This question requires 2 line graphs, one for each mesh size. It is sensible to use some way of discriminating the points for each set of data. A key is essential here.

(a) (i)

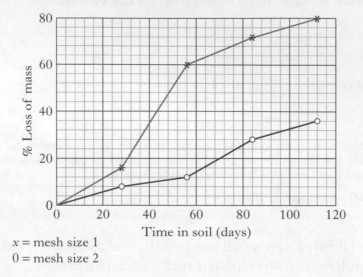

x = mesh size 1
0 = mesh size 2

[2] PS

HINT ▷ 1 mark for correct completion of each axis and 1 for plotting points.

(ii) 1

[1] PS

(b) one bag might be in a wetter area than other/some material might be sticking to the bags during weighing

[1] PS

(d) large/small mesh size increases/decreases the rate of loss of mass

large/small mesh size results in an overall greater/smaller loss of mass

[1] PS

10.

There are many areas in the Standard Grade course which lend themselves to a diagrammatic representation such as the sequence of events during mitosis. Using diagrams is a powerful way to help your revision of such areas.

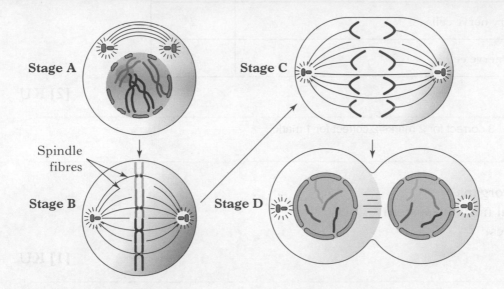

(a) B chromosomes/chromatid pairs line up along the middle of the cell/ equator

[1] KU

C chromatids/chromosomes pulled apart/separated/move to opposite ends of the cell

spindle fibres pull apart chromatids/chromosomes

[1] KU

(b) ensures no loss of genetic material/number of chromosomes is kept the same

[1] KU

HINT This is a very commonly asked question.

11. (*a*)

Description	Number
Organ which reacts to the stimulus	1
Motor nerve cell	4
Sensory nerve cell	2
Relay nerve cell	3

[2] KU

> HINT   3 correct for 2 marks/2 correct for 1 mark.

(*b*) sense organs
central nervous system
muscles

[1] KU

> HINT   All correct for 1 mark.

(*c*) tendons

ligaments

[2] KU

12.

 Students often find calculating cell sizes difficult through lack of practice with the basic unit of measurement which is the micrometre. Try to get a feel for the relative sizes of cells so that you can tell if your answer is "reasonable" or out by one or more decimal places.

(*a*) (i) 30 μm

[1] PS

HINT You are given the length as 0.03 mm so you simply have to multiply this by 1000 to get the answer in micrometres.

(ii) 7.5 μm

[1] PS

HINT You simply have to divide your answer to (i) by 4 to get this one. If you had got a wrong answer to (i) but did the calculation to (ii) correctly, you would not be penalised twice.

(iii) 6

[1] PS

(iv) 30 : 7.5 : 5

[1] PS

HINT Normally in answering this type of question, find the highest number that all numbers divide by without needing to use any decimals. Usually one number will reduce to 1. However, in this case, you cannot express this as a simple whole number ratio but make sure you preserve the order of the cells asked in the question.

(*b*) (i) Water moves [up/<u>down</u>] a concentration gradient as it enters a cell by osmosis. This is because the water is in [<u>higher</u>/lower] concentration outside relative to inside the cell. The cell will [shrink/<u>burst</u>] if this movement continues.

[1] KU

HINT Osmosis is a problem area for many candidates. Try to see it as a special case of diffusion rather than something quite separate. It involves a concentration gradient, as does diffusion, but also a semi-permeable membrane to separate the solutions. Water moves down the gradient, remember, so look where there is a higher relative concentration of water which will be at the "top" of the gradient.

All the correct answers are needed to get 1 mark.

(ii) oxygen/glucose/amino acids

[1] KU

13. (*a*)   sewage acts as a source of nutrient/food/energy for the bacteria.

[1] KU

(*b*)   bacteria perform aerobic respiration which needs oxygen

[1] KU

(*c*)   increased protein/energy content

[1] KU

(*d*)

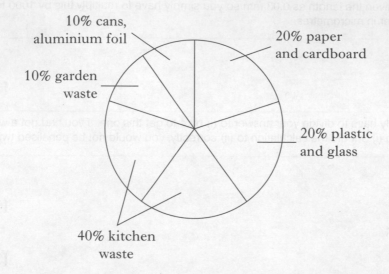

[2] PS

HINT	There are many ways to present data and a pie-chart is one you will quite possibly meet in a Standard Grade Examination. Basically, the circle is divided up into segments, or sometimes you have to construct these yourself. Remember you must include both the description and the data to be sure of scoring full marks. Make sure any lines you add to segments are very carefully drawn with a ruler.
	1 mark for correct plot of data and 1 mark for correct labelling.

14. (*a*)

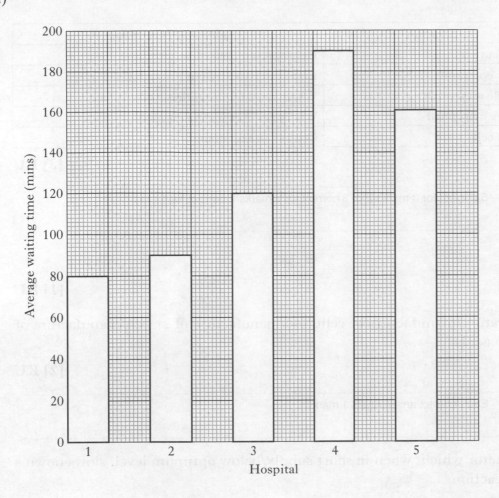

[2] PS

> **HINT** 1 mark for correct completion of each axis and 1 mark for correct drawing of the bars.

(*b*) 128 mins

[1] PS

> **HINT** With averages, add up the total (here 80 + 90 + 120 + 190 + 160 = 640) and then divide by the number (640 ÷ 5 = 128).

(*c*) 2 × /twice

[1] PS

15. (*a*)

Letter	Structure	Function/Description
A	Waxy cuticle	Helps prevent loss of water
B	Upper epidermis	Light can pass straight through this layer
C	Palisade mesophyll	Where most photosynthesis takes place
D	Spongy mesophyll	Where gas exchange takes place
E	Guard cell	Control size of stoma

[3] KU

> **HINT**   5 correct for 3 marks/4 or 3 correct for 2 marks/2 correct for 1 mark.

(*b*)   lignin

[1] KU

(*c*)   respiration/manufacture of cellulose/manufacture of starch/manufacture of amino acids

[2] KU

> **HINT**   Each correct answer gets 1 mark.

(*d*)   (i)   factor which, when in short supply/below optimum level, slows down a reaction

[1] KU

> **HINT**   The concept of limiting factors is a difficult one. Biological systems have a number of variables which impact on them and can slow them down if in short supply or below optimum level in some way. For example, temperature, light intensity and concentration of carbon dioxide are examples of variables which can affect the process of photosynthesis. The variable in shortest supply is the limiting factor.

(ii)   temperature/light intensity/carbon dioxide concentration

[1] KU

> **HINT**   Two correct answers are needed for 1 mark.

16. (*a*) A

[1] PS

(*b*) 7.5

[1] PS

> *HINT* Do not always assume a question on enzymes and pH refer to the human system. These enzymes come from a mammal other than a human. Notice how the activities of the three enzymes overlap at various places.

(*c*) 7.4 and 10.6

[1] PS

> *HINT* Always take time to read the scale carefully. Here there is no doubt of the answer but sometimes a little leeway is given and expressed as ±.

(*d*) 2 × /twice

[1] PS

PRACTICE PAPER B   WORKED ANSWERS

1. (*a*) (i) enzymes released from germinating seeds/amylase convert starch to sugar

[1] KU

  (ii) no competition from other micro-organisms
suitable temperature

[2] KU

(*b*) as temperature increases, enzymes involved in germination become more active
percentage germination will increase

[2] KU

(*c*) as milk sours bacteria respire anaerobically/without oxygen
breaking down lactose to form lactic acid

[2] KU

2. (*a*)

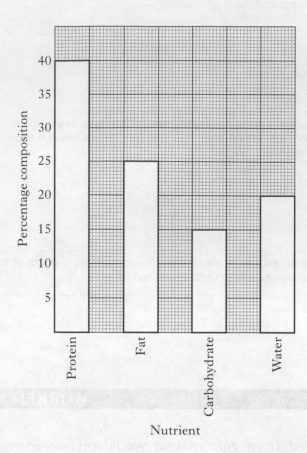

[2] PS

(*b*) 8 : 5 : 3 : 4

[1] PS

(*c*) total % for all four foods = 100
25% of this is fat

mass of fat in 280g of this plant = 25/100 × 280 = 70g

[1] PS

3.

 A variety of techniques could be chosen such as quadrat, netting or, as here, a pit-fall trap.

Pit-fall trap

1 Source of error   animals obtained may not be representative of the whole area being investigated

   How to reduce this   take many samples

2 Source of error   some trapped organisms may be eaten by birds

   How to reduce this   cover the opening to make it difficult for predators to see trapped organisms

[2] KU

HINT   Many experimental procedures and pieces of apparatus are themselves sources of errors in investigations. You should be aware of those you need to know for the Standard Grade and how you would minimise or eliminate each source.

HINT   Another possible source is that the trapped animals may actually eat each other and this can be prevented by including a preservative into the trap to kill the animals thus preserving all of them.

4. (*a*) 1   animals with increased strength

2   animals which give increased yields of milk

[1] PS

(*b*) nobody fully understood exactly which animals resemble their parents

[1] PS

(*c*) he repeated his experiments over many years

[1] PS

(*d*) the pea plant has a number of easily observable differences

[1] PS

(*e*) the parents were always true-breeding

[1] PS

(*f*) animal breeders can now direct the process of selecting animals with desirable features

[1] PS

(*g*) also allows the study of some of the processes involved in evolution

[1] PS

5.

Only by working through the past papers can you hope to become familiar with the different "styles" of questions. The same information can be asked in many different ways as shown in this question.

(a) (i)   C E

[1] KU

(ii)   A

[1] KU

(iii) B

[1] KU

(b)

Structure	Name	Function
A	Lacteal	To absorb digested fats
B	Capillary	Absorb digested carbohydrates and proteins/end products of digested carbohydrates and digested proteins/ glucose and amino acids

[2] KU

HINT You need to be careful in giving the function of the lacteal. Remember it doesn't just "absorb" digested food but specifically digested fats. Notice that several possible answers are possible for the function of the capillary but you must include the end-products of both carbohydrates and proteins and not just one.

6. (a) **Key**   ■ species A   ● species B   [3] PS

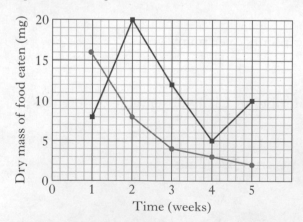

HINT Don't forget to include a key when you are drawing more than one set of data and don't plot values you don't have. For example, there is no "start" or "zero" value for the time.

One mark for completing axis and one for each correct plot.

(b) species B continually falls each week whereas species A goes up and down/ species A starts at a lower point than B but finishes at a higher value than B at the end of the investigation

[1] PS

(c) 8 mg

[1] PS

(d) fish B would continue to lose weight/die

[1] PS

7. (a)

	Number of eggs produced	Length of time of parental care	Protection given to eggs/ young	Chances of young developing
Fish	Many	Weeks	Almost none	Low
Mammal	Not many	Months	Protected inside mother's body	High

[2] KU

> **HINT** All four correct will get two marks. Three will get one mark but two or less correct will score zero.

(b)

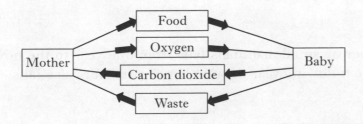

[2] KU

> **HINT** Four correct examples for two marks. Three or two correct for one mark.

8. (a) 35 − 5 = 30 mph

[1] PS

(b) 2 : 5 : 1 : 7

[1] PS

(c) hare

[1] PS

9. (*a*)

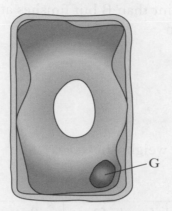

G

[1] KU

> **HINT** Draw lines without arrowheads and make sure they end exactly on the structure intended.
>
> Genes are located in the nucleus of the cell.

(*b*) the water concentration outside the cell is lower than that inside so water moves down the concentration gradient out of the cell

[1] KU

> **HINT** Make sure you explain why the water has moved rather than state what has happened. You need to show that you understand the concept of a "concentration gradient" to do this.

10. (*a*) total number of predators = 65

average is 65 ÷ 5 = 13

[1] PS

(*b*) when the number of small mammals falls from one year to the next in the same month, the following year the number of predators also falls

[1] PS

> **HINT** This question relates to the relationship between the number of a predator and its prey species. As one changes, it influences the number of the other.

(*c*) 4

[1] PS

(*d*) continue the sampling for a longer period/collect data over shorter time intervals instead of once a year

[1] PS

11. (*a*) (i) A

[1] KU

> **HINT** Remember that fat gives approximately twice as much energy as the same mass of either carbohydrate or protein.

(ii) $15 \times 20 = 300$ kJ/g

[1] PS

(b) growth/repair/movement/keep body warm/conduct nerve impulses

[1] KU

> **HINT** Any two for the mark.

(c) carbon, hydrogen and oxygen

[1] KU

(d) metabolism

[1] KU

12. (a) 10%

[1] PS

(b) 50% of 200 = 100

[1] PS

(c) 2 : 5

[1] PS

> **HINT** Make sure you preserve the order in answering ratios.

(d) this would increase the reliability of the results

[1] PS

13. (a)

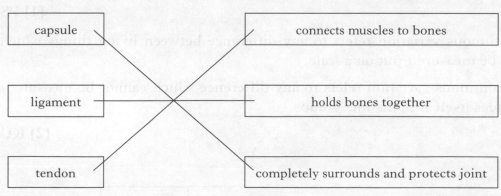

Part	Function
capsule	connects muscles to bones
ligament	holds bones together
tendon	completely surrounds and protects joint

[2] KU

> **HINT** Three correct for both marks, two correct for one mark. Anything less would get no award.

(b) muscles cannot push

when one muscle contracts, it needs another one with opposite action to straighten it

[2] KU

> **HINT** It is very important you reflect the number of marks in your answers. Here two marks are awarded for the explanation so make sure you give two distinct points.

(c) synovial fluid

reduces friction

[1] KU

14. (a)

Nutrient solution	Dry mass (mg)		
	Leaves	Roots	Total
A	2500	1400	3900
B	1814	1136	1950
C	2300	1200	3500
D	4000	3000	7000

[2] PS

> **HINT** Three correct answers for two marks and two correct for one mark.

(b) $3500 \div 20 = 175$ mg

[1] PS

(c) the dry mass of the leaves is always greater than the dry mass of the roots

[1] PS

(d) D

[1] PS

15. (a) continuous variation refers to any difference between living things which can be measured/put on a scale

discontinous variation refers to any difference which cannot be measured/divides itself into distinct groups

[2] KU

(b)

Example	Mutation	Selective breeding
Wheat for making bread has many sets of chromosomes and gives high yields	√	
Cauliflower has been produced from "ancient" cabbage plants over many generations		√
Hundreds of years of careful matings has resulted in Friesian cows which give very creamy milk		√
Disease-resistance in tomato plants	√	

[2] KU

> **HINT** When asked to insert a "tick" try to follow the instruction. Don't write in "yes", for example, or include crosses which, as here, you are not told to do. You may not be penalised for this but it is good examination practice to follow instructions.

(*c*) (i) change in the genes/genetic information/structure of the chromosomes/DNA/chromosome number

[1] KU

(ii) radiation/benzene/mustard gas

> **HINT** "Chemicals" is a little too vague here so be specific and name the mutagenic chemical.

[1] KU

(*d*) a cross looking at only one characteristic/parents differ in only one way with regard to the characteristic being studied

16. (*a*) 6000 ÷ 5 = 1200 N

[1] PS

(*b*)

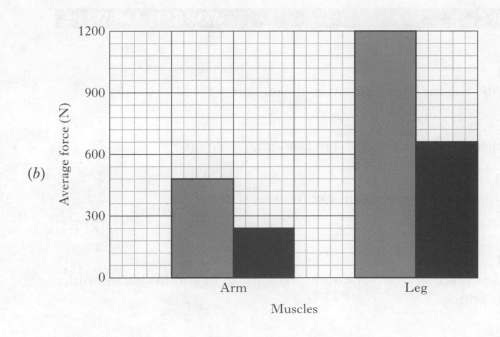

[2] PS

> **HINT** One mark for completing axis and one for correct plot.

(*c*) 480 ÷ 240 = 2

[1] PS

> **HINT** Notice you are not asked for the difference in the average forces which would be 240 N but how many time greater so the answer is twice as great.

(*d*) boys can apply more force in both muscles than girls
in both boys and girls, the leg muscles are about 2.5 times stronger than the arm muscles

[1] PS

(e) 400 + 1100 = 1500 N

[1] PS

17. (a) food developing new varieties/exploit existing plants not currently eaten

[1] KU

medicines such as drugs/antibiotics/cancer treatments

[1] KU

> **HINT** Here you need to go beyond just stating the use of a plant or plant product but add a little more to describe how it is useful.

(b) plant species may be used up at such a rate they cannot survive/climate change/ erosion/deforestation may cause their extinction

loss of genetic store/potential source of plant material for future/habitats/ food source for animals

[2] KU

PRACTICE PAPER C   WORKED ANSWERS

1. (a) Response: 1
keep their gills wet for breathing/avoid drying out

Response: 2
find food source/less energy wasted in random movement

Response: 3
need water to reproduce/keep skin wet for gas exchange

[2] KU

(b) Response: 1
breeding takes place in best conditions/avoiding cold conditions of winter when food is scarce

Response: 2
food supply is scarce/less energy is needed to survive winter

Response: 3
less liable to be eaten/less visible to predators

2 [KU]

> **HINT** There are a variety of common triggers for setting off rhythmical behaviour patterns in animals, such as changing daylength and/or changing temperatures. Usually, increasing daylength is associated with increasing temperature but they are two different triggers.

2. (*a*)

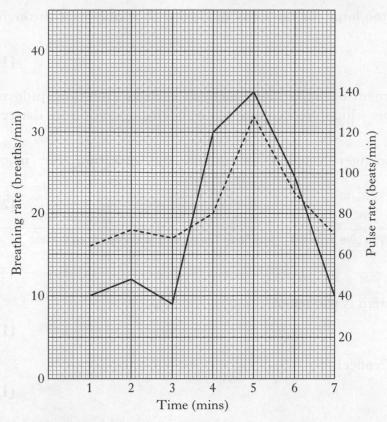

Key: Breathing rate ———   Pulse rate ------

[3] PS

 HINT   With a problem-solving question like this, which involves drawing two line graphs against different scales, it pays to take a lot of time to plot the points carefully.

(*b*) $10 \times 500 = 5000 \text{ cm}^3$

[1] PS

(*c*)   increase the size of the sample/use more pupils

[1] PS

(*d*)   3 – 4 minutes

[1] PS

3. (*a*)   1   stomach

2   small intestine

[1] PS

(*b*)   type of alcohol/how quickly the person is drinking/stomach full or empty/ age/ weight/ gender

[1] PS

HINT   Any two for one mark but, remember, no half marks for one correct answer!

(*c*)   the brain is still developing

[1] PS

(*d*) if the intake is too large for the body to cope with it may pass through the skin as sweat

[1] PS

(*e*) there is impairment of important functions such as attention, judgement and coordination which are necessary during the operation of dangerous machinery

you might endanger yourself or others by operating the machine improperly

[2] PS

> *HINT* Make sure you give two valid points here. The second follows on from the first.

(*f*) the immune system is weakened

[1] PS

(*g*) development of cancer

[1] PS

4. (*a*) (i)

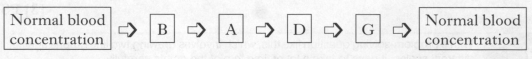

[1] KU

(ii) H

[1] KU

(*b*) Advantage   no longer need dialysis/free to move around/able to go on a normal diet/drink as much fluid as you want

Disadvantage   there may be problems with rejection/surgery is required to perform transplant/may require drugs constantly to prevent rejection/may not be easy to find a suitable donor/requires permission to have kidney removed donor

[2] KU

5. *(a)*

Concentration of sugar (g/l)	Percentage change in mass
0	+25
50	+20
100	−20
150	−20

[1] PS

HINT

Notice that the change can be + or −. To obtain the change, work out the increase/decrease and divide this by the initial mass then multiply by 100 as shown.
$(+0.7 \div 3.5) \times 100 = +20$ $(−0.7 \div 2.8) \times 100 = −25$
You must include the sign in each case.

(b)

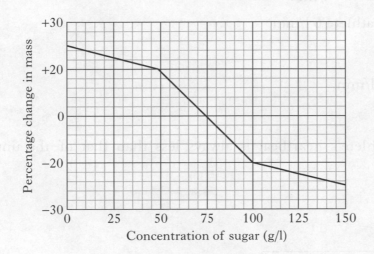

HINT

Always take time to plot line graphs carefully.

[2] PS

6. (a) (i) Oxygen demand cannot be met so cells start to respire anaerobically
This results in the build up of lactic acid which is painful

[2] KU

> *HINT* Notice in an "explanation" type question, it is not enough to state what is happening. You say why John had to stop. This is linked to the pain he will experience.

(b) pulse/breathing rates don't increase as much/recovery time is much shorter/
tolerance to the build-up of lactic acid increases

[1] KU

> *HINT* Any two correct responses for the mark.

(c) (i) trained athlete 7 l/min

untrained athlete 6 l/min

[1] PS

(ii) 10 − 9 = 1 l/min

[1] PS

(iii) trained athlete's heartbeat is always less than that of the untrained
athlete

[1] PS

7. (a)

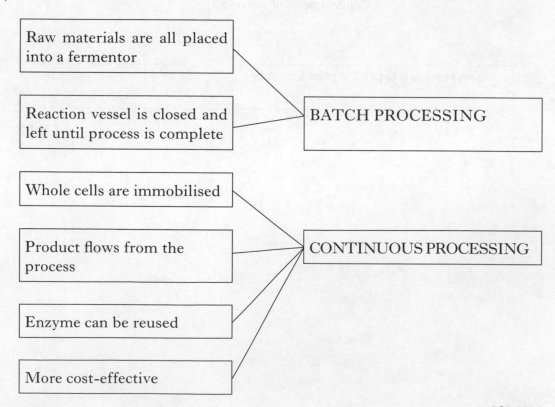

[2] KU

> **HINT** All five lines correct for two marks. Four or three for one mark.

(b) (i) remove any dissolved oxygen/kill any other micro-organisms/sterilise the solution

[1] KU

(ii) to prevent the loss of heat

[1] PS

(iii) to prevent gas exchange/keep oxygen out of the solution

[1] PS

(iv) test/show the presence of carbon dioxide

[1] KU

(v) set up the same apparatus using an equal quantity of dead yeast cells

[1] PS

> **HINT** Make sure you understand the need for control experiments in Biology.

8. (a)

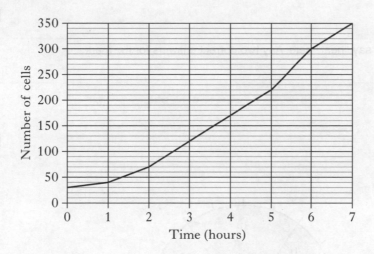

[2] PS

(b) 5–6

[1] PS

(c) number of living cells would fall dramatically

nutrients would be used up/toxins would accumulate

[1] PS

> **HINT** Both answers are required here to obtain the mark.

9. (a)

Statement	True	False	Correction
<u>Denitrifying</u> bacteria release nitrogen gas into the air from nitrates	√		
<u>Decomposers</u> convert nitrogen gas into nitrates		√	Nitrogen-fixing bacteria
<u>Nitrites</u> are taken up by plant roots to be made into protein		√	Nitrates
Bacteria convert <u>nitrates into nitrites</u>		√	nitrites into nitrates

[3] KU

> **HINT** The nitrogen cycle is always a problem for students, especially as some of the terms sound the same. Make sure you have this fully learned up with a good diagram to help.

(b) (i) bacteria/fungi

[1] KU

(ii) suitable temperature/oxygen/moisture

[1] KU

> **HINT** Don't just say "heat" here. Any two correct answers for the mark.

(c) root nodules

[1] KU

10. (a)

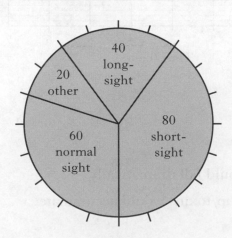

[1] PS

HINT There are twenty partially marked segments and since the total number was two hundred, each segment must therefore represent ten people. Make sure you label the pie-chart carefully. The segments can be in any order but usually try to follow the data. You could also shade/colour the segments and put a key to the side instead of writing the text but you must have the values as well.

(b) 200

[1] PS

(c) $(40 \div 200) \times 100 = 20$

[1] PS

(d) 4 : 2 : 1 : 3

[1] PS

HINT 80 : 40 : 20 : 60 is not the correct answer since you are instructed to give the answer as a simple whole number ratio so look for the highest number which can divide into all the figures; here it is 20.

(e) normal sight

[1] PS

(f) increase the sample size

[1] PS

HINT A favourite question!

11. (a) (i) peristalsis

[1] KU

(ii)

State of muscle	Result
A Contracted	Pushes food lump forward
B Relaxed	Allows lump of food to move easily

[1] KU

HINT Both answers required for the mark.

(iii) arrows should indicate food lump moving from A towards B

[1] KU

(b) mechanical action of contractions break up food into smaller particles/gives a bigger surface area for stomach enzymes to act upon/mixes up food with enzymes.

[2] KU

> **HINT** Any two for the two marks.

(c) liver

[1] KU

> **HINT** Remember bile is not an enzyme but it helps to break up fat, increasing surface area for enzymes to act on.

12. (a) May and September

[1] PS

> **HINT** Notice line graph hits 40 twice each year.

(b) 60

[1] PS

(c) January 2006

[1] PS

(d) cats   summer   dogs   winter

[2] PS

13. (a) cross will be Gg × gg

	G	g
g	Gg	gg
g	Gg	gg

(i) phenotypic ratio   1 green : 1 white

(ii) genotypic ratio   1 Gg : 1 Gg

[2] KU

> **HINT** It is always very wise to use a grid as shown to work out the answers, even if you are confident you know them!

(b) gg × gg

[1] PS

(c) no chance

[1] KU

> **HINT** GG × GG will not give rise to any white-flowered plants.

(d) ¾ of the plants would be green giving an expected number of

$(3 \times 76) \div 4 = 57$

[1] PS

(e) sample size may be too small/not all the seeds will germinate/some plants will die

[2] KU

HINT   Any two correct answers for the marks.

14. (a)

[1] KU

(b) (i)   people are living longer so more people have diabetes

[1] KU

(ii)   no issues of rejection/safer/no animals are involved/unlimited source

[2] KU

HINT   Any two correct responses for the marks.

(c) using lower temperature saves energy/clothes are not damaged at lower temperatures/enzymes are better at removing biological stains

[2] KU

HINT   Any two correct responses for the marks.

15. (a) companion cell C and sieve plate A

[1] KU

(b) guard cells

[1] KU

1. (*a*)

Structures	Inhalation	Exhalation
Intercostal muscles	Contracted	Relaxed
Ribcage	Moves up and out	Moves in and down
Lungs	Inflate	Deflate
Diaphragm	Contracts and moves down	Relaxes and moves up

[2] KU

> **HINT** All four answers are needed for the two marks. Three or two answers would gain one mark.

(*b*) dense blood supply/large surface area/thin-walled/moist

[1] KU

> **HINT** Any two for one mark.

(*c*) (i) A   cilia   B   mucus

[1] KU

> **HINT** Both answers needed for the mark.

(ii) bacteria/dust stick to the mucus
cilia sweep the mucus up the mouth for swallowing.

[2] KU

> **HINT** Notice that two points are needed here to explain fully the function of both structures.

(*d*) red blood cells

to carry oxygen

[2] KU

> **HINT** One mark for each answer.

2. (*a*) one-third / 1/3

[1] PS

(*b*) by closing one eye and allowing half their brains to go to sleep.

[1] PS

> **HINT** Both points are needed here to obtain the mark.

(*c*) they will die

[1] PS

(*d*) organising new memories and allowing brain to let go of random experiences

[1] PS

> **HINT** Both points are needed here to obtain the mark.

(*e*) feelings of hunger get switched off

we don't need as much food

[1] PS

> **HINT** Both points are needed here for the mark.

(*f*) being asleep means less likely to be eaten by predators

sense of sight does not work so well at night

[1] PS

> **HINT** Both points are needed here for the mark.

(*g*) impossible to know what is going on in somebody's head when they are sleeping

[1] PS

(*h*) animal or human has to be kept awake.

[1] PS

3. (a) (i) E
   (ii) B
   (iii) F
   (iv) A

[2] KU

> **HINT**   All four answers for two marks. Three or two for one mark.

(b) growing different crops each year in the same area of land

[1] KU

(c) A

C

[1] KU

> **HINT**   Both answers needed for the mark.

4. (a) A

water has entered the cell down the concentration gradient by osmosis causing the cell to burst.

[2] KU

> **HINT**   Try to give a full explanation to questions so that you obtain all the available marks.

(b) cells are able to exchange useful substances/get rid of waste materials/ examples such as movement of glucose/oxygen/carbon dioxide

[1] KU

5. (a) 5 + 8 + 12 + 13 + 12 + 12 = 62 millions

[1] PS

(b) $(5 \div 62) \times 100 = 8\%$

[1] PS

> **HINT**   It is perfectly acceptable to round the answer up to 8 here.

(c) 14 − 8 = 6 millions

[1] PS

(d) 15–29 years

[1] PS

(e) People born in 1900 did not survive as long as people born in 2005

[1] PS

6. (*a*)

Part	Name	Function(s)
A	Cerebrum	Memory, thought, intelligence, personality
B	Cerebellum	Balance/muscle coordination/posture/precise movements
C	Medulla	Unconscious functions/control of heartbeat/control of breathing

[2] KU

> **HINT** All four answers for two marks. Three or two answers for one mark.

(*b*) each canal is in a different plane/each canal is set at right-angles to the other

each movement will disturb the fluid in at least one canal and this will be interpreted by the brain

[2] KU

(*c*) each eye sees a slightly different picture

brain puts both pictures together to form one to judge how far away object is

[2] KU

7. (*a*)

Animal internal	Animal external	Explosive	Wind
4	2	3	1, 5

[1] KU

(*b*)

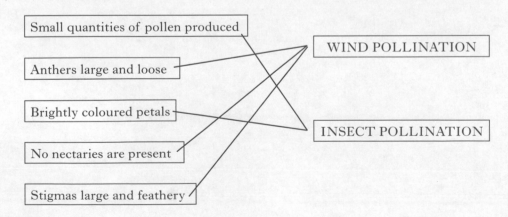

[2] KU

> **HINT** All five lines correct for two marks. Four or three lines for one mark.

(*c*) allows the best characteristics of two different plants to be combined

[1] KU

8. (a)

Facing	% cover
North	92
South	80
East	60
West	48

[2] PS

(b)

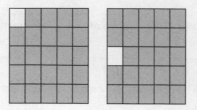

[1] PS

HINT > Any shading which leaves one box clear is acceptable. Two possible answers are shown.

(c) *pleurococcus* grows best/least when north/west facing

[1] PS

(d) by using more trees/increase the sample size/make sure the quadrats are spaced equally round the tree

[1] PS

9.

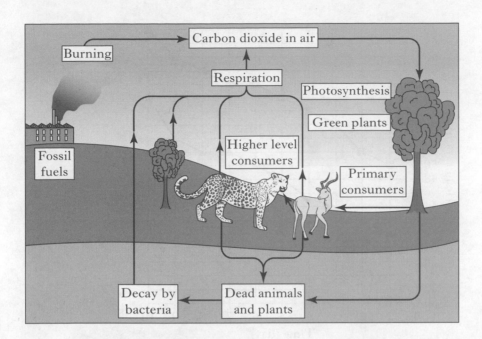

(*a*)

[2] KU

> **HINT** All four answers correct for two marks. Three or two for one mark.

(*b*) dead organisms would accumulate/not be broken down

[1] KU

10. (*a*)

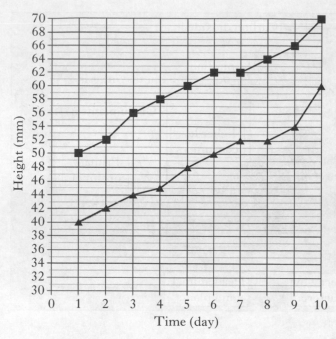

Key: Seedling A ■   Seedling B ▲

[3] PS

> **HINT** It is usually a good idea to indicate each plotted point as done here. Join points with a rule and point to point. One mark for each plot and one for completion of the *x*-axis.

(*b*) (i) Seedling A   increase was 70 − 50 = 20 mm

percentage increase is (20 ÷ 50) × 100 = 40

Seedling B   increase was 60 − 40 = 20 mm

percentage increase is (20 ÷ 40) × 100 = 50

[2] PS

(ii) fertilizer was inhibiting growth in some way

[1] PS

> **HINT** Do not always "expect" answers. Here the addition of the fertilizer seems to have inhibited growth of the seedlings.

(*c*) temperature/light intensity

[1] PS

(*d*) use a larger sample/more seedlings

continue to measure the heights for longer

[1] PS

11. (a)   the wolf population would increase

[1] KU

(b)   shows the changing numbers of organisms as you move up a food chain

[1] KU

(c)   have nitrogen-fixing bacteria attached to their roots/in root nodules

these bacteria can convert nitrogen into nitrates

[2] KU

(d)   (i)   Stage A   population is small and increasing slowly

Stage B   population is increasing and growing very rapidly

[2] KU

(ii)   they are the same

[1] KU

12. (a)   salt content of mammal 1 is half that of mammal 2

[1] PS

HINT ▷ Don't answer this type of question with "It is half" since this does not make clear to the examiner what "it" refers to. Be explicit in your answer. Also, notice an exact relationship is given as "half" rather than "less than".

(b)   mammal 2 is growing more rapidly than mammal 1/mammal 2 is larger than mammal 1

[1] PS

(c)   4 : 3

[1] PS

HINT ▷ This is the simplest whole number ratio and cannot be reduced further.

13. (*a*) Enzyme   A

Substrate   B

Product(s) C

[1] KU

> **HINT** All three answers are needed for the mark.

(*b*) pepsin will not act on any other substrate/only acts on protein

[1] KU

(*c*) temperature at which the enzyme works best/is most active

[1] KU

14. (*a*) Male pupils   175 cm to 190 cm

Female pupils   154 cm to 164 cm

[2] PS

> **HINT** If asked for a range, make sure you start from the lowest value and end with the highest value.

(*b*)

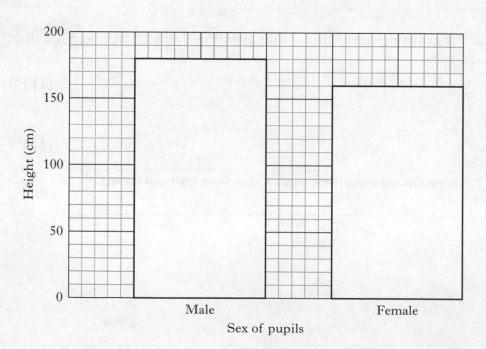

[2] PS

> **HINT** Notice the *x*-axis requires a full labelling. Make sure the bars are spaced apart and the same width and use up at least 50% of the grid.

(c) always use the same measuring device/same technique for measuring/same person to carry out the measurement/have pupils remove their shoes before being measured.

[1] PS

15. (a)

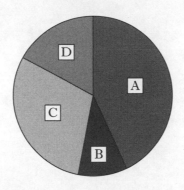

[1] PS

(b) 200

[1] PS

> *HINT*   86 + 20 + 60 + 34 = 200

(c) only a small sample of plant material was used

[1] PS

(d) Cell type A 48%   Cell type B 17%

[2] PS

> *HINT*   Simply divide each percentage by 2 since the total was 200.

(e) 48 : 10 : 30 : 17
    A   B   C   D

[1] PS

> *HINT*   This is the lowest whole number ratio you can obtain.

16. water is essential for the transport of sperms/land-dwelling animals have no water to carry sperms to the eggs

internal fertilisation ensures the sperms, once inside the female's body, are in a watery environment/can swim towards the egg